Monographs on soil survey

Soil classification for soil survey

B. E. BUTLER

CLARENDON PRESS · OXFORD
1980

Oxford University Press, Walton Street, Oxford OX2 6DP

OXFORD LONDON GLASGOW
NEW YORK TORONTO MELBOURNE WELLINGTON
KUALA LUMPUR SINGAPORE JAKARTA HONG KONG TOKYO
DELHI BOMBAY CALCUTTA MADRAS KARACHI
NAIROBI DAR ES SALAAM CAPE TOWN

Published in the United States by Oxford University Press, New York

British Library Cataloguing in Publication Data
Butler, B E
Soil classification for soil survey. – (Monographs on soil survey).
1. Soils – Classification
2. Soil-surveys
I. Title II. Series
631.4'4 S592.16 79-41141

ISBN 0-19-854510-X

Set by Hope Services, Clifton Hampden
and printed in Great Britain by
Lowe & Brydone Printers Ltd, Thetford, Norfolk

Dedicated to John K. Taylor, devoted investigator and worker in Australian soils: who for some 30 years as my Chief actively encouraged me to re-examine old ways and ever to explore new avenues of soil enquiry, especially those related to soil survey. His encouragement and discussion I gratefully acknowledge, though knowing that he may not agree with some of the conclusions I have reached here.

B. E. Butler

Editors' Foreword

The grouping of phenomena into classes that can be talked about or used as the basis for further generalization is an activity so fundamental that in most instances we do not realize that we are doing it. The classification of soils, that is the recognition and definition of groups of soil individuals with properties in common, has proved more difficult and controversial than most. Much is claimed; little is proved. The best schemes of classification have provided useful frameworks for generalizing the results of research and experience so that identifying the class to which a soil individual belongs automatically tells us something about its properties and behaviour. The worst have divided soil on irrelevant properties without proof that the groups created share any useful properties. Some of them require more information about a soil individual to put it into its class than could ever be gained from knowing to which class it belongs. One voice urges the student to classify, another tells him not to bother since crop yields or management practices are poorly correlated with soil classes; neither shows him how to judge.

Bruce Butler, a past President and Honorary Member of the Australian Society of Soil Science, is probably best known outside Australia for his elucidation of the control of soil development by landscape processes, as expressed in the concepts of prior streams and K-cycles. He has been a soil surveyor throughout his career and has faced the problems of finding usable and useful soil classifications.

Here he approaches the problem of classification head-on. Why do we classify? What do we hope to get out of it? To what extent can existing classifications meet this need? Should we use an international scheme of classification or create a local one? These questions must be asked. He introduces the principles of the 'Taxonomic Chop' and the 'Taxonomic Hiatus'. Finally he tackles the crucial problem – what the soil surveyor should do – and offers a tentative procedure, which the surveyor may follow as it stands, or use as a framework for his own analysis of his particular project.

P.H.T. Beckett
R. Webster
V.C. Robertson

Acknowledgements

The author is pleased to acknowledge substantial assistance with the manuscript by P. H. T. Beckett and R. Webster. M. Mulcahy and G. D. Hubble also read the manuscript and gave the benefit of their views. Stephen Sheppard drew the free-hand diagrams.

The author also acknowledges the kind permission to quote their copyright material given by the following publishers, authors, and organizations: the Education and Publication Branch, U.S. Soil Conservation Service to quote from *Soils and men* and from *Soil taxonomy*; the Editorial Branch, F.A.O. and E. Wegman, Unesco to quote from *Soil map of the world–Vol. 1*; Rellim Technical Publications to quote from *Handbook of Australian soils*; George Allen and Unwin to quote from de Sigmond *Principles of soil science*; E. A. FitzPatrick and Oliver and Boyd publishers, to quote from *Pedology*; the Minister of Supply and Services, Canada for permission to quote from *The system of soil classification for Canada*; Oxford University Press to quote from B. W. Avery in *Journal of Soil Science*; the Council for Scientific and Industrial Research Organization to quote from R. Webster and B. E. Butler in *Australian Journal of Soil Research*; and the author gratefully acknowledges permission from the Director Consejo Superior de Investigaciones Cientificas to quote from *The soils of Europe* by the late W. L. Kubiena.

Contents

1. Introduction: the nature of soil survey

This is a book about soil classification and how the soil surveyor creates a list of soil classes that will comprise all the soil in the area he is to map.

It will be simplest to start by describing what the soil surveyor does in a soil survey and why it is usually necessary for him to produce a soil classification; then to discuss the problems that in some circumstances make it difficult if not impossible to produce a useful classification; and finally to propose a procedure that, in its essentials, can be applied to any survey and (we hope) minimize the risk of producing a classification that is useless. At the very least he would wish to know before he begins mapping that the soil of his area is in fact classifiable and mappable and that the classes and mapping units he proposes are relevant to the particular problems of management or planning that the survey is to help solve.

Throughout the book we can discuss the broader classes of a national or regional classification to which the surveyor may relate his own local soil classes so that he can draw on the experience of managing similar soils elsewhere.

Soil survey is one of the basic technologies of soil science and it has a long and respected record. But even though the procedures for recognizing soil groups and determining the positions of soil boundaries on the ground are well established and basic to soil survey, there has been remarkably little examination of their theoretical basis. Although the *Soil survey manual* (U.S.D.A. 1951) has been and still is a standard reference in its field and the parent of many national handbooks, most of it is taken up in the definition of classes, the formalization of soil and site descriptions, and the techniques of making a map. It offers little discussion of the basic philosophic questions that are implicit in so many of the decisions the soil surveyor has to make. *Soil taxonomy* (U.S.D.A. 1975) is subtitled 'a basic system of classification for making and interpreting soil surveys' but it offers little discussion on the philosophy of soil survey. According to Blackburn (1960) the Russian conventions of soil survey also fail to relate the craft of soil mapping to the philosophic principles of classification.

Soil survey

A soil map should have a purpose; for example, to help solve a particular problem or to provide information on which to plan the location of a particular land use or the development of a region. It should either provide information on those properties of the soil that affect an intended use of the soil itself, or serve as an index to a separate library or store containing such information.

If there is no-one in the survey area who thinks he has a soil problem it is unlikely that a group of eager users of soil information will spring up merely because a soil map has been produced. This would seem axiomatic, but a surprising number of soil maps has been produced for no better reason than an unfocused if benevolent interest on the part of the central government or equivalent. So the surveyor's first responsibility is to confirm that a knowledge of the distribution of different kinds of soil within the area of interest, which is what he would contribute if he were to produce a soil map, is in fact required by one or more significant groups of users.

Assume that there is a specific problem. The soil surveyor will try to ascertain which soil properties or aspects of soil behaviour are causing the problem or appear to be related to it. In his context these are the *useful* or *relevant properties* of the local soil. There will be no point in mapping the whole area in intricate detail before discovering, for example, that the problem had arisen only from excess soil acidity, which could have been mapped by itself in much less time.

At the same time the surveyor must recognize his own limitations. He is not trying to solve the problem—that is a matter for the agronomist, engineer, or veterinary surgeon. What he can do is to ascertain the distribution in his area of the soil properties that affect the problem and to record and present them using a classification map and its legend. His role is to *predict*, that is to provide a means or a tool with which the problem-solver can inform himself about the relevant soil conditions at all sites in the area *without having to go there to see*. How he is to do this is the subject of this and other volumes in this series.

This concept of prediction is important. Webster (1977) has discussed it also. The point is that there is only one way of ascertaining, with total confidence, the state or the attributes of the soil at a site and that is to go there to measure or observe them. The average user of the land, or planner of competing land uses, is unable or unwilling to do this; the soil surveyor is brought in to do it for him, and more specifically to employ all the skills he has to provide soil information about the whole area of interest without the cost of visiting every point in it. In short, the surveyor is being asked to provide a tool (his map, memoir, key etc.) from which the user can inform himself about the soil conditions at any site without having to go there. This is what is meant here by prediction. In the same way the success of the surveyor's work will be judged by the extent to which his predictions about the soil at sites of interest prove to be correct. Terms such as the *predictive power* of a classification or map, or the *predictability* of soil attributes in an area covered by a classification or map, will recur frequently in this book. Indeed the author is convinced that several of the most elegant classifications have very little predictive power outside the limited areas from which their component classes have been derived.

Ideally the surveyor would map the values of all relevant properties all over the area. In practice, this is rarely done. In expensive irrigation schemes, for example, it may be economic to map the variations in exchangeable sodium, soluble-salt content, clay content etc. over all the irrigation area. Even in this case the survey is usually simplified by creating classes in these properties; for example, by representing exchangeable sodium as 0-7½%, 7½-15%, > 15% of the cation exchange capacity, conductivity of the saturation extract as 0-2, 2-4, 4-8, 8-16, > 16 mmhos/cm, and clay content as 0-18%, 18-35%, 35-60%, > 60%. The surveyor then need not look for and map all values, but only the class limits. This is not unreasonable; if the farmer or the irrigation engineer, for example, cannot conceive of applying gypsum in finer grades than 500, 1000, or 1500 kg/ha, there will be no point in mapping the corresponding values of exchangeable sodium more finely.

Relevant properties are often expensive or laborious to determine, and many of them can only be determined in the laboratory, so that the surveyor has to collect samples, have them analysed, and then interpolate class boundaries between his observation sites some time afterwards. His boundaries would probably be better if he were able to interpolate them between successive observation sites on the evidence of local changes in vegetation, relief, or surface appearance.

Usually he looks for soil properties that may or may not be useful in themselves but are easily assessed in the field, preferably by hand or eye; these include the morphological properties of colour, texture, rootedness, gleying, or so on. He then defines soil classes from some of these properties. The classes may be defined by the limiting values of their morphological properties (definitional classes) or by their central concepts (typological classes) (Gilmour 1962). The surveyor exerts himself to produce a list of classes (a *classification*) that is complete enough to identify all the soil in the area. When the surveyor *classifies* the soil of an area, he erects a classification or a system of classes; when he *identifies* a soil entity he decides to which of these classes it belongs, or *allocates* it to a class, or gives it its class name.

The soil mantle is a continuum that extends over a range of conditions and its properties vary accordingly. It often lacks sharp discontinuities, so it is not always clear what exactly are the objects or *entities* that are to be identified. Much energy has been devoted to discussing this, and terms such as profile, pedon, or worse have been much bandied about, but the discussion has had remarkably little effect on what soil surveyors actually do. Let us say that in most circumstances a column of soil of 1 m in diameter is large enough to comprise adequately most of the short-range variability characteristic of any soil area without being too large and encompassing a range of soil variation that common sense requires to be divided into two or more soil classes. In this case, it will be convenient and pragmatically justifiable to treat such 1 m columns as soil individuals or soil *entities*. We shall not be pedantic either about the more general terms 'soil', 'the soil', 'soils', and 'a soil' of which the common usages are inevitably confused, and may refer to soil as a material ('soil is a good rooting medium' or 'the soil in my garden'), to the whole range of soil variability in an area ('the soils of Berkshire are mostly calcareous' or 'most of the soil in Berkshire is calcareous'), or to indicate soils of particular but unspecified kinds or profiles ('some of the soils in the Canberra area are acidic'; 'this research was mainly applied to soils in the Western Tablelands'). The reader will find no difficulty in distinguishing between soil as a material, all soil material, all soil profiles, some of all kinds or types of soil material, or soil profiles, from their context.

There are situations in which important soil properties vary rhythmically with periodicities of 1–20 m, where perhaps the units are too finely recurrent to justify the creation of two classes but the contrast in behaviour is too great to be encompassed within one class. The surveyor will probably be inclined to map such an area as a defined combination of two or more defined subclasses.

Given a list of soil classes, defined on their morphological properties, any entity may be identified on its values for these *definitive* properties. If, as is hopefully assumed, other soil properties are associated or correlated with these, then the classes will also be found to group soil entities according to their values of *associated properties* (e.g. it may be that all soils with mor humus (a morphological property) have surface pH less than 5). If, in addition, the classes group soil entities according to their values of relevant properties, then the classes will usefully separate soil entities according to the differences in management that they require. It cannot be assumed, though it often is, that the definitive properties and the useful properties of a set of soil classes are associated.

In a limited area it is often possible to identify the members of a soil class on less than the full range of its definitive properties. For example, if a group of deep, sandy, and acid soils with a textural *B* horizon that contains high exchangeable aluminium also comprises all the red soils in an area, this class may be distinguished from all other classes from the red colour alone. This single character may then be used in an *identification key* to separate the members of this class from those of all other classes, on the assumption that this *keying property* is uniquely associated with its other definitive properties, and in the hope that it is also associated with its other useful properties. A *keying property* should be unambiguous. All members of the class must exhibit the character; members of all other classes should not, unless the classes are distinguished on more than one character, e.g. the four classes defined on the possession of two attributes: *AB*, *A*-, *B*-, --. In this case there are two taxonomic chops (p. 19) employing the criteria *A*/ not *A* and *B*/ not *B*. The class limits could also be numerical, e.g. $A > 100/A \leqslant 100$. At the same time a key must be local. If the survey area were enlarged till it included two kinds of red soil, the first group would no longer be uniquely distinguished on its redness and the key would break down.

Once he has developed a classification—his gallery of conceptual soil classes—the surveyor can start to map them. Two points quickly emerge. First, their pattern of distribution is usually intricate and the boundaries that separate different classes are correspondingly tortuous. He will have to simplify them to some extent and the boundaries on his maps will be simpler than those in nature, so the areas or *mapping units* that they enclose will not be completely pure but will contain inliers of other classes.

Secondly, if the surveyor can locate these boundaries only by repeated soil borings he will never be able to make enough borings to locate correctly all of even the simplified boundaries he wishes to record. Usually he tries to find features of the landscape (e.g. changes in relief, natural vegetation, colour of soil surface that he can see continuously on the land surface) that correspond to the soil differences he wishes to represent, and he then uses these to interpolate the boundaries between his borings. Indeed, unless he is mapping on a large scale, his decision on how to split up the whole soil continuum into soil classes may be influenced by the extent to which the classes in alternative classification systems correspond to mappable surface features. Where two or more classes occur in irresolvable intricacy he will probably map them together as a *complex* or *compound unit* especially if the differences between them are of little significance for the purpose of the survey.

The final result then is a map made up of simple or compound mapping units, accompanied by a *legend* which records the soil class(es) to be expected within each mapping unit and explains any symbols or inscriptions, and a *memoir* which states definitive characters, perhaps offers an identification key, and summarizes the associated characters of the classes. This is offered to the user who requested the survey or to others so that they can put names to every kind of soil in the survey area.

The users may wish to do this for several reasons:

(a) They wish to see as much soil data as possible arranged in one orderly compilation: a tidiness of structure in the organization of their thoughts;

(b) They wish to put a *name* to any soil entity so that it may be discussed and related to other soil entities already identified;

(c) They believe that once a soil entity has been identified it will be possible to *predict* things about it by associating it with the fund of experience and information belonging to its class, and, with modification, to other similar classes. *In short, they expect to be able to predict certain things of the identified soil which they could not have done before going through the exercise of identifying it.*

The surveyor's classification relates specifically to the local realm of soil experience. Probably soils with the same definitive features will occur immediately outside the survey area, and these can be given the same class name, but the surveyor should not expect his local classes to extend over a great area. Nor, if he finds distant soils that match the definitive features of his classes, should he assume that they necessarily share the same associated or useful properties. Nevertheless, it will usually be possible to group them into higher more comprehensive classes, and then into yet higher. Some national classifications have five or six hierarchical levels, in which the classes at each higher level become progressively more comprehensive in concept and wider in their geographical extent.

2. Problems: the concept of orderliness in the soil population

Chapter 1 outlined what the soil surveyor does, and what he hopes to achieve by it. Specifically he is attempting to provide a means or tool by which the user of the land may predict the soil conditions at any site in his area of interest. Now we must consider the problems and the difference between the ideal and reality.

Most soil surveys are done because there is a *problem*: there is some soil-use proposition in the area on which a soil attribute has a large influence. The soil survey is done to show where particular management options should be exercised to make the best use of the area or to avoid expensive failures.

This need to establish and analyse the relationship between the problem and soil attributes in the project area must be emphasized. In its absence, many a soil survey after its publication has lain unused either because there was no soil factor to 'the problem' or because its association with the classification and legend had not been worked out and demonstrated. By working out the soil associations of 'the problem' during the planning and development of a soil survey, it is possible to assess how effective the survey may be at solving the problem, before the major operation of soil mapping is undertaken. It is quite possible that in some cases the mapping will not be done because it has become apparent that there is a better way of handling the problem.

In coping with 'the problem' in relation to soil surveys there has been a tendency in the past to assume that the relationship between soils and problems which have been established in one locality may be 'exported' and used in distant places with the same success and efficiency as at their place of origin. Indeed national and world soil-classification systems are developed to exploit this supposed advantage. But when we reflect that the connection on the soil side depends on a few keying properties (usually morphological) and the connection on the problem side depends on one or two associated edaphic properties, it will be realized that the connection is often tenuous and unreliable.

'The problem' requires serious investigation. In particular it is necessary to ascertain that it has a soil factor and that there is an association between that soil factor and mappable soil characteristics in the survey area. In the past this has not been clearly recognized as either the soil surveyor's or the agronomist's responsibility, and often the result has been that the job has not been properly done.

In some cases a soil survey has no particular problem but is made to provide a general stocktaking of the soils of the project area. Undoubtedly such a survey can provide interesting information about the main soil variables and how they relate to each other and to the landscape; it may also be helpful in broad resource planning. Often the boundaries coincide neither with those required for detailed planning nor with the boundaries of management options for some particular problem. Indeed, an unfocused survey may produce too many boundaries, and it will require comprehensive investigations to show which are the relevant ones for particular problems.

We shall start then from the stage when the surveyor has identified the problem that he is to help solve and has made at least a mental list of the soil properties that are relevant to the problem. If the survey is to be part of a general stocktaking of a natural resource, then he has established a list of the soil properties that at least determine the main anticipated uses of the land. It is his intention that his survey will predict the probable values of these relevant soil properties at any site of interest, or their predominant values within any management area in the survey area.

The surveyor's soil classification is a means to this end. It would be extremely useful if every soil entity belonged to a soil class, and if, for all the members of each class, the soil properties relevant to the problem lay within defined ranges. If so, experience could be passed on from one entity to another in the same class; merely to identify the class to which a soil belonged would specify its main properties. To achieve this the surveyor would try to match his classes to the natural 'modalities' of his area; there is no point in defining classes for ranges of property values that do not exist, nor in defining classes that do not correspond to the differences which the farmer or other users perceive to be important.

The soil surveyor is also concerned with the correlations which he can establish to assist him to locate soil boundaries, so he is interested in the correlations between soil profiles and the topographic features of slope, shape, stoniness, and various features of land use and vegetation, or indeed the correlation between soil profile and the appearance of the soil surface and its vegetation cover on air photographs or on the line-scanning printouts of remote sensors. In general then, the soil surveyor's main technical repertoire is his ability to develop and utilize correlations. He seeks to correlate useful properties with features of the soil profile and to relate them to a shortlist of keying attributes. He then allocates unknown soil entities to their classes on the basis of their *keying* attributes and uses the fact of their class identity as the basis on which he can *predict* what the status of other attributes will be. This prediction is an extrapolation from his original correlation.

The concept of the soil key was briefly introduced on p. 5, and may now be elaborated a little. We have already established that one of the purposes of making a soil classification system is to enable unknown soil entities to be identified and hence acquire a name, be better understood, and have predictions made about them. The identification is achieved by searching through the classification to find the lowest class into which the unknown soil will fit. This can be done laboriously and with possible uncertainty by matching the soil description to class descriptions, but it may be done much more quickly and confidently if the classification system is supported by a soil key.

The maker of the soil-classification system probably has a more complete awareness of the associations and variations of properties throughout a given population of soils than anyone else and is therefore usually best able to produce the sequence of yes/no questions which will bring the unknown soil to its proper class in the system. Such a key will be monothetic (i.e. a necessary and sufficient condition for the membership of a class is that an entity possesses the attributes specified in the key). The success of the key at putting entities into classes depends on a firm association between the keying characteristics of a soil class and its other definitive characteristics.

The growing trend toward the development of keys as part of soil-classification systems is motivated by the need for precise and quick identifications of new soil entities. A good key will achieve this, but in itself can achieve no more. There is no certainty as to what the associated properties of the created classes will be nor that they will be consistent from place to place. Since our interest in the study and use of soils usually lies among their associated and useful properties rather than their keying properties, the service provided by keys is often of doubtful value. Their facility is deceptive. Since they are usually bifurcating in operation, they can be applied to any population of soil entities but they will not necessarily produce groups of soil entities that show consistency in their useful properties for any soil population other than the one for which they were developed.

This leads us to the main problems of classification in soil survey. Notably it reminds us that there were implicit assumptions at every stage in the exposition of the 'normal' soil survey in Chapter 1. If the survey is to be of any substantial help these assumptions must have been met; namely that

(a) the individual entities belonging to each class are more or less continuous so that most classes can be mapped in mapping units that are both coherent and not too impure;
(b) a reasonable proportion of the boundaries to be drawn possess external expression to guide their delineation;
(c) the definitive properties of the classes are either themselves relevant to the land uses being considered or else are associated with useful or relevant properties;
(d) the relevant properties are in fact relevant and between them account for most of the management problems that gave rise to the survey.

All these assumptions relate to what may be called the '*orderliness*' of the soil mantle in the survey area. This is a difficult but useful concept. If the local soil is 'orderly' then classification and mapping are likely to be straightforward and may be useful; if it is not then they are unlikely to be useful and may be very difficult.

This chapter now discusses and brings into the open various facets of this concept of orderliness.

Attributes

All soils have a great many properties and attributes and, while this enhances their likely points of interest and use, it increases the problem of generalizing about them and of classifying them. It is convenient to think of soil information in terms of the three groups of soil attributes shown as the three parts of Fig. 2.1. The properties of the soil as we see and handle it comprise the most common facet. This is the descriptive facet of soil morphology as described horizon by horizon. These attributes include the colour, texture, structure, and consistency of soil material; the arrangement and thickness of the soil horizons; micro-morphological attributes; and some descriptive analytical data.

The second facet comprises the soil-forming processes by which the

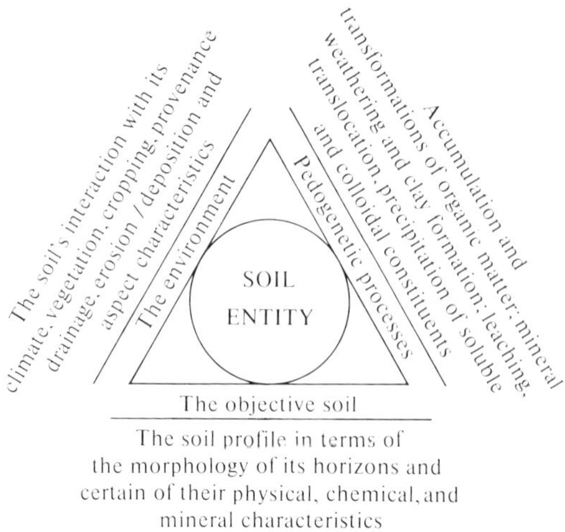

Fig. 2.1. The many characteristics and attributes of a soil expressed around three facets. A soil entity comprises a complete set of data, but we often use only a limited number of these, whichever are accessible at the time, or of use in our study, or whichever will identify a soil, or will relate to a land-use problem. Notwithstanding these *partial* interests, we know that the soil is present as a whole assembly of data; but we also know that identification cannot involve more than a few attributes.

soil has been formed from its parent material, and those properties that reflect them; e.g. soluble salts, organic matter, sesquioxides, clay etc. The pedogenetic processes are largely deduced from the inter-horizon relationships of a soil profile. It is assumed that a soil's morphology clearly indicates the soil processes which established and now sustain it. Particular studies have confirmed the validity of this proposition in particular cases, but not every soil profile examined in a soil survey can be subject to such studies and the few verified conclusions are extrapolated very widely on their associations with soil descriptive features.

The third facet of soil data is that dealing with the soil's environment such as its siting or position in the landscape, its climate and drainage, and especially the organisms of the ecosystem with which it is associated. The environmental conditions are believed to have a direct and comprehensive control over the pedogenetic processes.

It would generally be agreed that all of the characteristics of these facets of a soil are of interest to the understanding of soils or in their use for agriculture or engineering.

However, the very difficulty of obtaining some items of information about a soil entity compared with the ease of obtaining other information inclines the surveyor to use the latter as a stand-in for the former. He is encouraged by the hypothesis that all soil characteristics are interrelated and that soil morphology is created by pedological processes which are determined by factors of the environment; i.e. like situations always produce like soils. Specific studies have confirmed the association between morphology, soil process, and soil environment in specific circumstances, but there have been too few such demonstrations to justify the use of morphologic characteristics as a currency to transact all soil information.

The use of substitute characteristics in soil studies and soil classification is widespread. *Soil taxonomy* (U.S.D.A. 1975) uses selected morphological attributes as differentiae for the higher categories of its system because of their pedogenetic significance. Kubiena (1953) uses both morphology alone and morphology interpreted as pedogenetic processes in his classification. The *World soil map* (Dudal 1974) uses morphological criteria in its soil classification which it claims are pedogenetic in significance. Muir (1969) assumes a complete association between chosen morphologic attributes and the dynamic soil characteristics.

These assumptions are not unusual but often there is inadequate proof of the associations on which they depend.

The distinction of classes

Let us imagine that the soil surveyor has familiarized himself with the soils of his area and erected a soil classification. He now wants to find out how far his classification separates and groups soils on attributes in which he is interested because of their pedologic or utilitarian interest. Let us suppose that one 'attribute of interest' is per cent base saturation. By carefully unbiased sampling the surveyor chooses a number of sites in his area; at each site he identifies the soil class

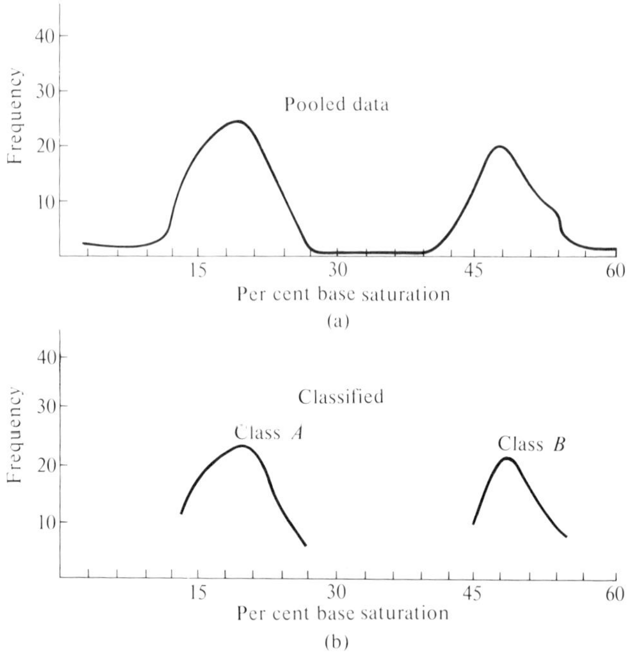

Fig. 2.2. In the ideal case the frequency distribution of an attribute of a population of soils shows distinct and separate modes. The whole population is plotted in (a) and the two classes *A* and *B* are plotted separately in (b). By chance or good judgement the two classes coincide with the separate modes.

and analyses a soil sample from a defined depth for its per cent base saturation. Fig. 2.2(a) presents a histogram, or a frequency distribution graph, for the whole set of data.

A frequency distribution curve usually indicates the preponderance of some attribute values over others. A high and peaky curve indicates that attribute values fall almost entirely within a specific narrow range. A broad low distribution curve shows a wide spread of values. Two populations of the peaky kind can be distinguished confidently even when the difference between them is not great, but two populations with broad distributions cannot easily be separated.

Fig. 2.2(a), with two clearly separated peaks (or two modal groups) one of high values and another of low values, shows a high level of organization in the data with regard to base saturation. Fig. 2.2(b) presents separate frequency distribution plots for class *A* and *B* entities in the sample. There is a clear distinction in base-saturation values between class *A* and class *B*; class *A* is always low in base saturation and class *B* is always high and each is very distinctive, so that the classification is highly successful. If any soil entity in this area is identified as belonging to class *A*, we could predict its per cent base saturation with confidence, especially when compared with class *B* soils.

Remember that, as a general rule, soils are not identified as belonging to class *A* or class *B* on the basis of their base saturation but on the morphological attributes which comprised the key for identifying these classes. Thus the gain in predictability of base saturation is an advantage from the classification and is not a purely mechanical result of sorting entities on base saturation itself. The system works because there is a correlation between base saturation and the one or two soil attributes used in the keying of these classes.

The theory of differentiae

We may now look at a less perfect example. Suppose the surveyor has established that the 'attribute of greatest relevance' is the P_2O_5-sorption at 0–10 cm depth. Fig. 2.3(a) is a frequency-distribution plot of the P_2O_5-sorption values of all soil entities in an unbiased sample.

At first sight this is not encouraging since the distribution curve is level over the whole range and no sorption values are more likely than any others, apart from those near the ends of the range. As far as this attribute alone is concerned there is no suggestion of orderliness. So the next option is to plot class *A* and class *B* entities separately to see if the separation reveals any orderliness which was not apparent in the data as a whole.

Fig. 2.3(b) shows the sorption values of entities in class *A* and class *B*. Curve *A* is indistinguishable from curve *B* and both show the same frequency distributions as the pooled data—on sorption status they are one. The classification has failed to introduce any order into the distribution of P_2O_5-sorption values; there is no advantage in separating class *A* from class *B* and the classification makes no contribution to predictions of the attribute of interest. The soil surveyor may abandon

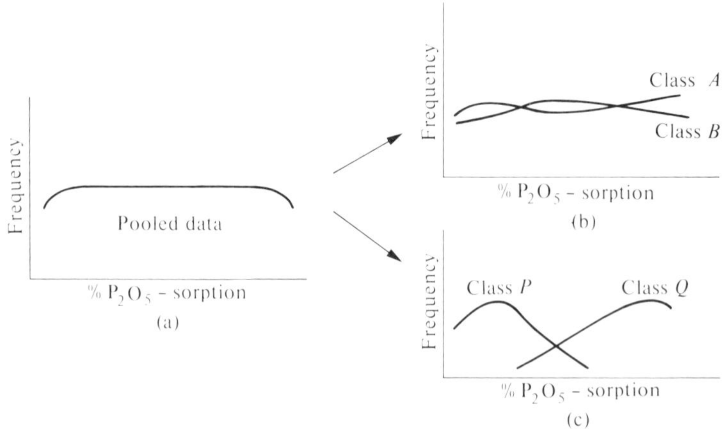

Fig. 2.3. Frequency distributions in a more ambiguous case. The pooled data (a) shows no characteristic orderliness: the first attempt at classification (b) makes no improvement on this, but the partial separation of the curves achieved by the second classification (c) shows the small but definite increase in orderliness associated with it.

his efforts at classification at this stage. However, he is typically indefatigable and believes that some classification schemes, some 'taxonomic chops', are more effective than others in imposing or revealing order in the soil data. He may choose to use an exotic classification based on some other soil landscape or he may look through his own collected data for other attributes which seem to correlate with the attribute of interest; he may also examine attributes which he infers from general soil theory to be important. Whatever choice he makes, he will use his chosen differential to create a new class *P* and a new class *Q*. Fig. 2.3(c) presents their separate frequency distributions for P_2O_5-sorption values. In this case, the distributions are distinguishable although they overlap. This degree of orderliness enables the surveyor to make some approximate predictions about the P_2O_5-sorption of any soil entity which he has allocated to classes *P* and *Q* within the confidence limits of the sorption status of each class. It is apparent that these are wide although they could be real.

Once again the second aspect of orderliness—the association of attributes—has brought itself into the analysis. The associations of attributes need to be examined thoroughly. There is a general tendency for soil scientists to subdivide populations of soils, and to identify the classes they create, on the basis of a set of attributes of one kind, while requiring that the classes so created are significantly different in terms of a set of different attributes.

These ideas can be explored further using scatter diagrams. Suppose that two attributes *X* and *Y* have been measured on a set of soil entities. Their values can be plotted against each other on a graph so that each entity will appear as a point the position of which is determined by the values of the two attributes. The graph as a whole shows the relationship between one attribute and the other for the entities measured. Fig. 2.4 shows several such scatter diagrams. It also shows several different kinds of orderliness or lack of it, and we shall now look at the implications of each.

Orderliness may appear as clustering or clumping; the plotted points may form distinct groups that are separated from one another by empty areas as in Figs. 2.4(b) and 2.4(d). Alternatively, orderliness may show itself in a single obliquely elongated cluster in which the points are concentrated near an oblique line as in Fig. 2.4(a), or points may fall close to a strongly curved line although this is unusual for soil. A scatter diagram that shows none of these characters lacks orderliness (Fig. 2.4(c)).

When the scattering is obliquely elongated as in Fig. 2.4(a), the two attributes are said to be 'correlated'; i.e. they are interdependent and any division of one attribute scale will tend to divide the population on the other to produce a distribution somewhat like that in Fig. 2.3(c). Thus a division of *X* at K in Fig. 2.4(a) will produce two groups in which the observed ranges of *Y* will be less than in the whole population. If *X* were a morphological attribute of no direct relevance (e.g. colour value between 0 and 10 cm depth) and *Y* were an attribute of interest (e.g. citrate-soluble P_2O_5 in the same depth range), then a classification based on colour could lead to reasonable predictions of soil phosphate, and the closer the points lie to the straight line the more precise will be such predictions. In fact, if the attribute *X* were measured, the values of *Y* would usually be predicted by regression without recourse to classification. But in soil survey practice it is not possible in the field to do more than assign values on very coarse scales to many of the morphological properties of the soil, and so classification is more appropriate.

Consider next Fig. 2.4(b) in which there are two clusters, one above and to the right of the other. Here a division of the *X* scale anywhere in the region of *M* will serve to distinguish two groups, each

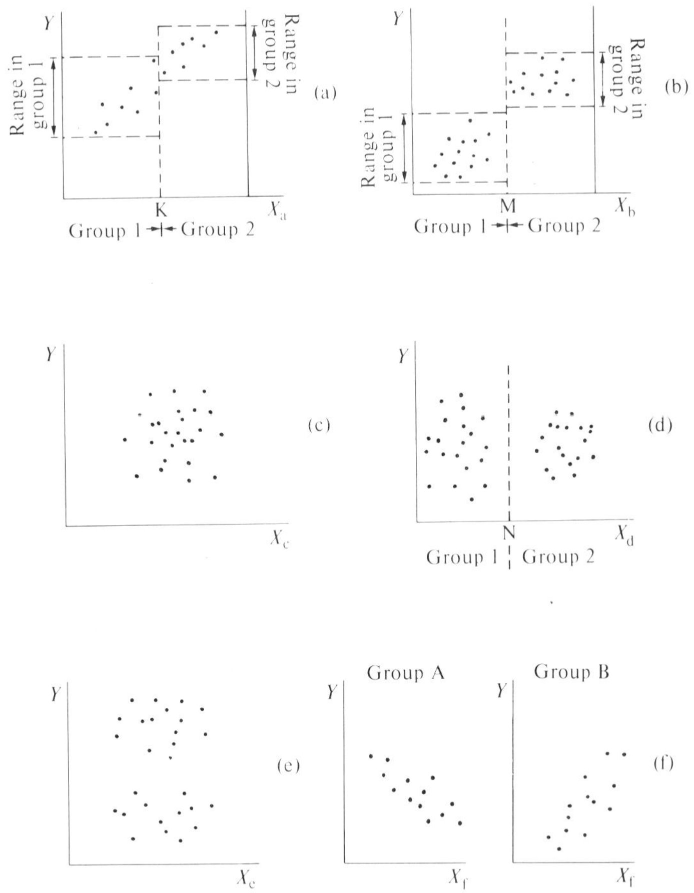

Fig. 2.4. Orderliness expressed in the association of two attributes as shown by scatter diagrams on which each point represents the X and Y values of one soil entity.

of which would have predicted values of *Y* lying within a reduced and definable range. Values of *Y* could be predicted by regression, but the predictions would not necessarily be any more precise than those achieved by classification.

The lack of orderliness in Fig. 2.4(c) has already been mentioned. If *X* represents soil colour and *Y* citrate-soluble P_2O_5, then there is no prospect of predicting P_2O_5 if we know only the soil colour. However, clustering alone without correlation will not allow profitable predictions either. In Fig. 2.4(d) there are two quite distinct clusters. A critical value in the vicinity of *N* will serve well to distinguish the two groups. However, since *Y* is not related to *X*, the range of *Y* is the same for the two groups and, if the aim is to predict *Y*, the classification is a failure. Alternatively the population might be distributed as in Fig. 2.4(e) with strong clustering on the *Y* axis. If there is no correlation between *X* and *Y* and no corresponding clustering on the *X* axis, a knowledge of *X* will not serve to identify the clusters and thereby cannot predict *Y*. Some other character should be sought for the purpose if *Y* cannot be observed directly.

Finally, we should note that even when two attributes appear to be uncorrelated as in Fig. 2.4(c), it may be possible to divide the population on a third attribute, say *Z*, so that the resulting groups exhibit distinct correlations between *X* and *Y* within each group produced. Thus from a Figure like Fig. 2.4(c) we might produce Fig. 2.4(f). Identification of the soil class, whether *A* or *B*, will not itself allow worthwhile prediction of *Y*, but it will enable *Y* to be predicted from *X*. Here again is an orderliness and, once it is recognized, classification will be useful.

The taxonomic chop

Separation of classes on arbitrary values of characters is known as dissection. We shall call the decision to make a division at any particular character and the choice of its limiting values the *taxonomic chop*. It is the executive aspect of a classification scheme and, when the machinery of a classification scheme embodies several taxonomic chops, they constitute a *key*. Any classification that uses this principle, and this includes many of the soil-survey classification schemes outlined in Chapter 4, should use taxonomic chops that make use of whatever order is present within the population of soils for which it is intended. The classifier should choose those chops of the accessible characters that will produce a classification with the greatest predictive power for other and preferably useful characters. Not all characters are equally good for this, nor are all values of the most satisfactory characters equally suitable as chop points. Returning to Fig. 2.4, the division of X at K in the top graph is a taxonomic chop. The character X is well chosen; the value or values at which it is divided are less important since almost any position for the chop within the observed range of X will enable Y to be predicted better. The values can be chosen for convenience or technological need. Dissection on X in Fig. 2.4(b) seems even more appropriate. However, to be useful the chop point must now lie in the vicinity of M. Taxonomic chops may also be applied to the X variables in Fig. 2.4(c) and 2.4(d); in the latter the position for the chop is obvious enough, but in neither case will they be of any benefit in predicting Y.

Generalization

The idea of prediction as the prime aim of a soil classification has been present throughout the text so far. This has been deliberate; prediction and the associated ideas of economy in sampling and experimental effort must be paramount in any professionally conducted soil survey. However, prediction is not the only reason for classification; there is also the need for generalization. Thus a soil surveyor may have measurements of the soil properties at a large number of sites. In discussing them with other people, especially non-technical people, he will usually want to assign the sites to just a few groups so that he can present a broad picture of each. It will be easier to talk about the groups and to remember them if the groups have names. The soil surveyor will often find it convenient to talk in general terms of acid and alkaline soils, shallow and deep soils, and so on, even when he has precise measurements of the pH and depth of a number of soil entities. For this purpose Fig. 2.4(a) could be split profitably at K, and Fig. 2.4(b) at M without regard to the effect on Y. Similarly one dissection of the X axis at N in Fig. 2.4(d) is worthwhile and some division of Fig. 2.4(c) may reasonably be contemplated. These are generalizations of convenience but, if several characters of each soil entity are measured, a large survey will produce such a large body of data that they must be condensed if they are to be simultaneously comprehensible. This raises new possibilities for the taxonomic chop and is the reason for much of the recent research on the application

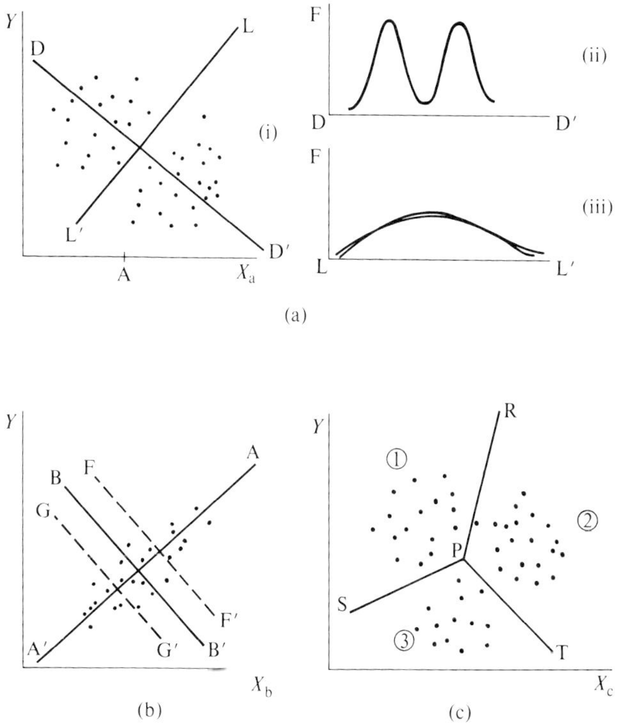

Fig. 2.5. Orderliness expressed in scatter diagrams of attributes X and Y.

of multivariate analysis to soil systematics (Webster 1977).

Consider first Fig. 2.5(a)i in which there is very evident orderliness with two distinct clusters. A division could be made on the X axis at say A, but this would leave some members of the upper left cluster in the lower right class. The cutting point could be moved to the right until all members of the upper left cluster were in their 'correct' class, but then the class would contain points from the lower right cluster. The problem remains whatever cutting value is chosen, and the same applies to the Y axis. However, it is possible to choose a line LL' which will pass between the two clusters and separate them neatly. This represents a taxonomic chop on an axis such as DD' at right angles to it. The lines LL' and DD' are linear functions of X and Y both of which must be known. The line DD' may be regarded as a discriminant axis (see Webster 1977); such divisions are the basis of discriminant analysis. The frequency distributions of the two clusters on DD' will then appear as in Fig. 2.5(a)ii. Almost any other taxonomic chop will result in a greater degree of overlap. The worst possible chop would be by DD' on line LL', which will produce two groups of which the frequency distributions on LL' would coincide as in Fig. 2.5(a)iii.

Alternatively, the orderliness may be manifest as a single obliquely elongated cluster. In Fig. 2.4(a) the chop was placed on the X axis only. If information about Y exists and is to be taken into account, then again a linear function of X and Y is needed. For example, in Fig. 2.5(b) the most sensible solution is to place the chop through the distribution at right angles to the major axis AA', e.g. by the line BB' or by the two lines FF' and GG' if three groups are desired instead of two. All other divisions would be less desirable and the worst possible chop would again be by the major axis itself or lines parallel to it.

The situation is somewhat more complex if there are more than two clusters. That illustrated in Fig. 2.5(c) is common enough. Dissection orthogonal to the X or Y axes would be unsatisfactory; a vertical chop through P would separate clusters one and two but split cluster three, and there is no horizontal line that will neatly separate clusters two and three. Nor is there a prominent major axis along which divisions can be made. However, three regions can easily be created by the lines RP, SP, and TP, each of which is a linear function of X and Y and may be chosen to lie at right angles to the lines joining the centroids of the clusters.

Finally, if there is no orderliness, as in Fig. 2.4(e), then if the population is to be classified for convenience, there will be no point in cutting it except in a direction orthogonal to the measurement axes.

It will be evident that the classifications that have been judged desirable are ones in which the groups created make use of and represent any orderliness present in the population. In general, there will be less dispersion within such groups than within the same number of classes in any other classification of that population. It should also be clear that suitable taxonomic chops cannot be chosen without reference to the population being classified. Although bureaucratic rules or exotic classifications may fortuitously coincide with the structure locally present, usually they will be less than the best and can at worst make nonsense.

Keys

When several taxonomic chops are applied in succession they constitute a key. As far as possible characters will be chosen for the key because they reveal orderliness rather than conceal it. They may be either single attributes or combinations of attributes, although in practice single attributes are more simple.

In a similar way the values of the chop points will be chosen to take maximum advantage of any gaps or zones of low population density in a distribution.

Essentially the soil key enables new soil entities to be allocated to the most appropriate classes of a classification. What does it achieve? In the first place, allocation of any entity to a class loses information. A precise measurement of X for that entity is replaced by the range of values encompassed by the class to which it is allocated. This is the price paid for the ability to say something precise about some other attribute Y not used in the key. In other words the aim is prediction as before. Ideally a key involves taxonomic chops on one or more identifying attributes associated with other attributes that are less easy or more expensive to discern. In soil survey the keying attributes are usually morphological, whereas the relevant attributes that are ultimately of interest are related to soil performance or to aspects of pedogenesis. For example, in Figs. 2.4(a) and 2.4(b) X is the keying attribute; once a soil entity has been allocated on it, Y can be predicted more precisely than it could be otherwise and the key has been useful.

On the other hand, in Figs. 2.4(c), 2.4(d), and 2.4(e) it is still perfectly possible to allocate any entity to its class on its values of X. The price paid is the same–loss of information about X–but the gain is nil. Even when an entity has been allocated to its class, it is still not possible to say anything more about its value of Y. Classes of this sort have no predictive value–they are *key-only* classes. Nothing can be said about members of the classes except that they possess the keying attributes–a fact already known which may not in itself be of interest. Such classifications contain a lot of 'noise' (i.e. variations that are unrelated to the keying criteria) and are in a sense failures.

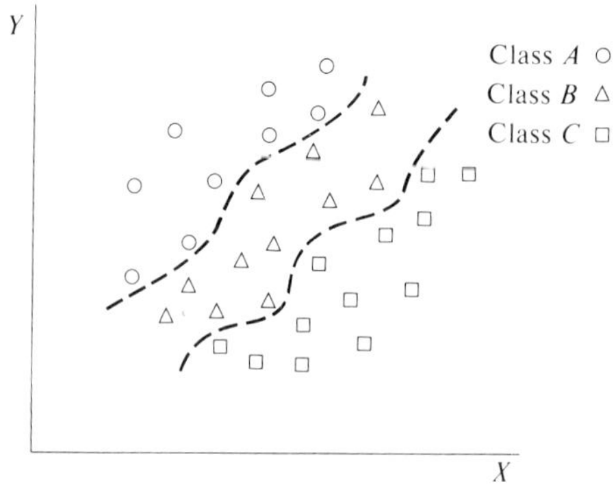

Fig. 2.6. The ideal effect of a classification resolving the apparent disorder in a population of soils. The soil population is represented as a plot of property Y on property X, which shows no orderliness for the data pool as a whole: X could not be predicted from Y. But when classes A, B, and C are plotted separately they each show a separate orderliness, and the prospect of predicting X from Y when class is taken into account.

Evidence of order

Much of the ethos of soil survey harks back to the early Russian work with its notions of natural bodies in equilibrium with their environment. Implicit in this is the assumption of a high degree of orderliness in which soil properties are strongly interrelated. Many classification schemes and their keys are based on this assumption. Reality is often different.

Fig. 2.6 illustrates the general assumption that classification resolves the seemingly wide-ranging disorder of a soil population into segments of order. The soil entities have been identified into classes A, B, and C, and are plotted in terms of their characteristics X and Y. It is apparent that this classification has created order in a population which had very little appearance or order.

Fig. 2.7 presents an actual case based on a sample data set of 111 sites covering some 400 hectares (Webster and Butler 1976) at Ginninderra near Canberra.* The soil entities are plotted in terms of

* The Figure is copied here by kind permission of the Editor in Chief, Council for Scientific and Industrial Research Organization, Australia.

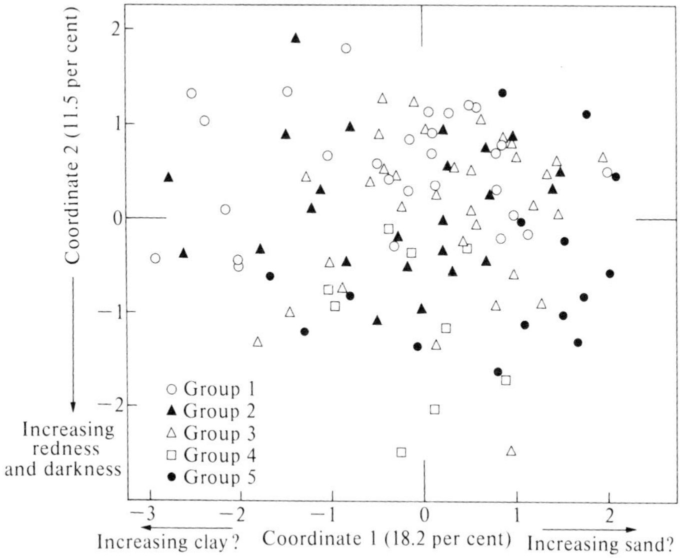

Fig. 2.7. Plot of actual soil classes from Ginninderra plotted on the first two principal coordinates of the data pool. The indications are of a real but low level of orderliness among the soil properties associated with the coordinates (Webster and Butler 1976).

their first and second principal coordinates calculated on the basis of their morphological and chemical properties. Each entity is identified as one of the five classes of a conventional classification scheme based on soil morphology. Clearly there is little order in the data as a whole, and division into classes does not enhance the appearance of order much. In this case the low degree of orderliness is borne out by other studies which are referred to below.

The association between attributes

Some of the situations discussed above are illustrated in Fig. 2.8. Each circle represents one soil entity. The eight radii in each circle represent eight (a very small proportion) of the characteristics of each soil. The characteristics are the same in each circle for each of the like numbered radii, and they should be thought of as including morphologic, pedogenetic, and environmental characteristics. The length of each indicates

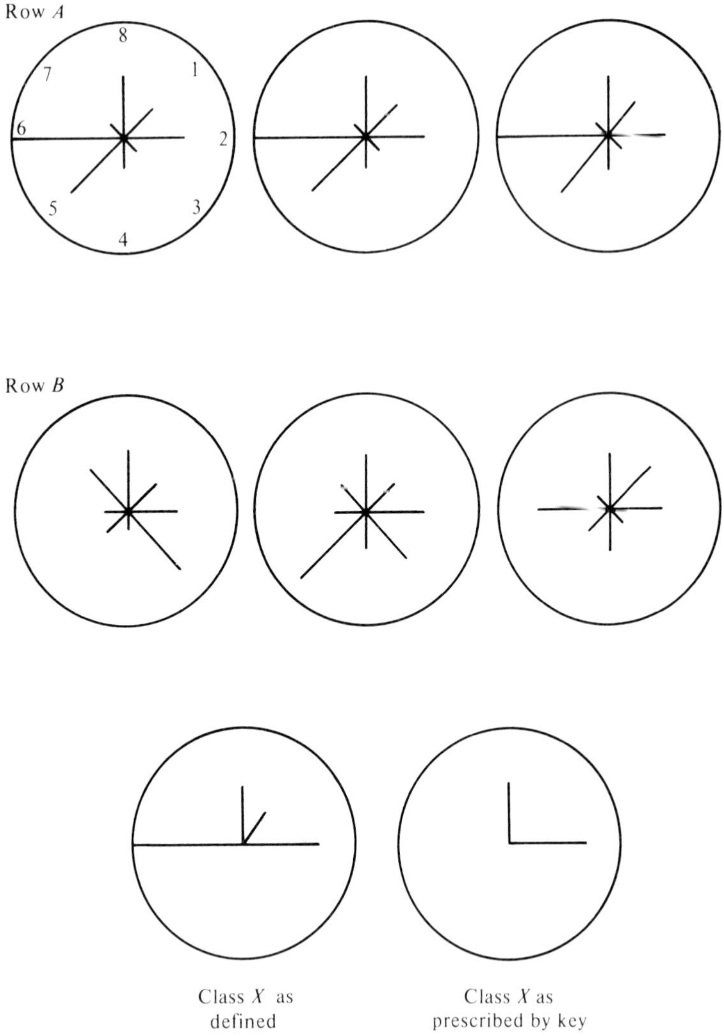

Fig. 2.8. A demonstration that when the same key is applied to two soil data pools, the groupings achieved may not be the same. Class *X* is defined on attributes 1, 2, 6, and 8, but since these are uniquely associated in the area for which the classification was created, the

the magnitude of the characteristic in each soil. Assume that a sample set of soils has been studied and a classification system properly drawn up for it with defined classes and a key for their quick identification. The bottom left of Fig. 2.8 shows the definition of class *X* and the bottom right shows the key for identifying class *X*. We see that class *X* is described in terms of characteristics one, two, six, and eight, and it is keyed by characteristics two and eight only as these are uniquely characteristic of the class within the limits of the area in which the classification was developed.

The key is then applied to a number of soil entities from the same area and row *A* shows the three soils allocated to class *X*. They have characteristics two and eight in common because in this area characteristics one and six are correlated with characteristics two and eight, and the key was developed on this basis. But the soil samples show even more agreement among themselves than this; the values of characteristics three, four, five, and seven are also matched. This free trade-off of *other* associated properties is the hoped-for result of every classification exercise. In this case, not only was the classification system correctly constructed in its classes and its key, but the sample set on which the classes were based properly represented the soil population as a whole, and the soil surveyor succeeded in framing classes in which the definitive and other properties are highly correlated. Not all classifications achieve this. In the Ginninderra study cited above, only 11–13 per cent of the total variance of soil characteristics was associated with the soil classes of the conventional classification.

In other circumstances the operation of the key would also allocate the three soils in row *B* of Fig. 2.8 to class *X*. In this row characteristics two and eight are common as they must be, but characteristic

---

class is keyed only on attributes 2 and 8. The entities of row *A* are drawn from the data pool on which the key and the specifications of class *X* were based, and hence they all show concordance in the keying characteristics 2 and 8, and also for properties 1 and 6 which are defined for the class but not used in the key: properties 3, 4, 5, and 7 also conform, this being the assumed ideal condition for soils, but perhaps rarely attained. The entities in row *B* drawn from a different soil population are correctly keyed to class *X* on attributes 2 and 8, but show only limited agreement with its other definitive attributes (1 and 6) and very poor agreement with each other or row *A* on attributes 3, 5, and 7.

six is mismatched, as are characteristics three, four, five, and seven. This could easily happen either if the classification and the key were based on a sample set of soils that were not representative of the area as a whole in their nondefinitive properties (e.g. when the soils of the area show a high but unsuspected variability in soil molybdenum, from benign to toxic levels, uncorrelated with any observable soil features), or more commonly because the soil surveyor has taken a classification and key derived from one region (in which it behaved as in row *A*) and applied it to another region. For example, early reports on the soils of Africa often report large areas of chernozems in the Southern Sudan, that are in fact vertisolic soils keyed into the chernozems on their depth of dark soil and their horizon of secondary calcium carbonate. Clearly the results of keying are very different in row *A* and row *B*, and yet the operation of keying would have seemed very similar in the two cases. A surveyor whose mastery of the key, particularly if complicated, represents the summit of his professional skill is unlikely to appreciate that the difference is important. The apparent precision of the keying operation has wide appeal among soil classifiers. Most keys operate on a bifurcating principle and give unequivocal identifications of any unknown soil entities that are referred to them. They have no built-in indication that other characteristics of these entities are associated, although such characteristics are often more important to the user than the keying properties which are matched. The situation where the soil entities in a class are found to have only their keying characteristics in common is very neat and tidy but quite useless. Since a classification system is 'good' according to the reliable *associated* properties found in its classes, we may consider that a system has failed when further entities submitted to it fail to have these associations, notwithstanding the uniformity of class imposed by the key. So soil surveyors should certainly attempt to construct classifications and keys, but they must not assume that the results are profitable. They must demonstrate that some other properties of interest can be predicted for the entities once they have been identified by the key.

Exotic or national classifications

As soil classes are recruited from ever increasing areas, the associations between their attributes become ever more diffuse. A limited range of primary minerals in one region in which, for example, phosphate and iron are to some degree associated may give rise to a set of useful associations between soil colour and citrate-soluble–P_2O_5 in that region, its form varying between morphologically defined soil-drainage classes. There is no reason whatever why there should be similar associations in another region of similar lithology but different mineralogy. The wider the area, the greater the chance that it will comprise significant but unrecognized variations of this kind. It is useful to realize that the level of orderliness in a soil population is likely to be greatest for an area of a few square kilometres and will diminish as the area of interest increases. When the data of several soil populations are merged–as when several soil-survey classifications are merged to make a national soil classification–the orderliness of the enlarged population is reduced still further and the predictive value of the soil classes is correspondingly reduced. The orderliness of soil classes is at its lowest level when this merging tendency is carried to the ultimate of a World soil classification.

Soils in the landscape

Much of soil classification and soil survey is concerned with soils as areas of land. Areas of land have physical extent in the three primary directions; they have juxtaposition and a surface to the atmosphere; and that surface has shape, texture, and surficial accretions including living organisms. Areas of land vary in all of their characteristics; the variation is of both major and minor degree and of local-repetitive or linear-secular kinds.

The soil surveyor stands in the midst of his problem–the landscape (Fig. 2.9) in which he wishes to draw boundaries to separate areas of one kind from other different areas. He will either use the landscape boundaries which he sees (if he can prove they also divide classes of soils) or, failing that, he will have to 'dig out' his soil boundaries by making a considerable number of soil examinations.

One of the main concerns of the soil surveyor is to establish a correlation between his soil classes and the classes of landscape in his project area. Soil theory postulates that the relief and drainage of the landscape are factors in soil formation, so most operators assume that the desired correlation exists. But the acceptance of a general rule is not good enough. If the soil surveyor looks to landscape boundaries for the location of his soil-class boundaries, an imperfect or partial association between landscape and soil boundaries will

Fig. 2.9. The surveyor can see boundaries in the landscape but which can he justify in soil-profile terms? The changes in relief and vegetation which he can see must be checked with the 1 metre soil auger to see what similarity in soil profile there is within topographically similar areas, and how different the soil is between topographically different areas.

greatly reduce the accuracy of his soil map. Bleeker and Speight (1978) have investigated some instances of very weak associations between these elements in two localities in Papua New Guinea. There is, in general, too little information available on the correlation between soil types and landscape subdivisions in specific cases. Clearly the association is too important to be presumed; it should be tested in any soil-survey project.

We shall consider first the soil surveyor's usage of landscape data and then consider the reconciliation of soil classes and landscape elements. The landscape in any given scene generally offers some well marked contrasts. As well as the contrasts between major elements, the observer who looks closely and critically will also find meso- and even micro-levels of variation in the landscape. Variation is in characteristics such as form, aspect, evenness, texture, and surficial accretions. At these several levels of variation he can see or interpose boundaries between contrasting areas and possibly see similar areas recur in some sort of orderly pattern. At the lowest level the surveyor can recognize unit areas of uniformity which, at a minimum scale of definition of about a hectare, are called 'sites' by Clarke and Beckett (1971) and 'land units' by Christian and Stewart (1968). These are the basic units of land in detailed soil survey and the primary soil-type areas of the soil surveyor. His first concern is to recognize them and then to investigate their soil characteristics and check the consistency in soil characteristics over a site and between similar sites. Cline (1963) says that the soil surveyor sees the boundaries on the land first and then examines the character of the soil. He claims that there is an association between soil boundaries and landscape boundaries, that every skilled soil surveyor is subconsciously aware of this association in his working area, and that soil survey at any but the largest scale would be impossible without it. With a choice of alternative sets of soil classes, the surveyor will be predisposed to choose those that fit the landscape.

General experience has shown that local soil variations in texture, colour, presence of accessory materials like lime and sesquioxide concretions, and changes in colour mottlings and soil wetness are often associated with topographic changes, although there is not always a one for one correspondance. These changes may be associated with geological erosion and the associated deposition on hillsides, the exposure of different parent materials, and variations in site drainage. These local soil variations account for most of the soil surveyor's conclusions on boundaries, because other soil variations due to climate and the associated vegetation vary so slowly with distance that the corresponding soil boundaries are unreal and not easily distinguished.

Taking the classification from the country

Taking the soil boundaries from the landscape boundaries is practicable in addition to filling the requirement (U.S.D.A. 1951) that the classification must fit the landscape. It also seems necessary because, if soil survey boundaries traverse apparently uniform areas and do *not* follow boundaries or obvious changes in the landscape, the map would not commend itself to the practical user. This generating of soil classes to match the visible discontinuities in the landscape we may call 'taking the soil classification from the country'.

The amount of soil variation in a unit area of landscape is likely to vary from one unit to another. At some sites it may conform to what can be accepted within one primary class or series; in other cases it may be too great for that and the area will need either to be designated as a complex of series or subdivided on the evidence of frequent augering or soil pits. This distinction between areas of one soil class and areas occupied by complexes of soil classes emphasizes the essential difference between soil classes and landscape classes. When mapping at medium to small scale the surveyor often attempts to recognize 'patterning' in landscape and soil type. Instead of looking for unit landscape areas or sites of a size to comprise one series only, the surveyor looks for areas occupied by a recurrent pattern of several kinds of sites and for the repetition of any such pattern in different parts of his project area. The soil types within a pattern or complex may be very diverse in character; it is their repeated occurrence in propinquity which is the basis for generalizing them into a complex mapping unit. The patterns are easier to see on aerial photographs than on the ground, and mapping the patterns on aerial photographs is a common means of generalizing the soil array of an area. A pattern is not a taxonomic generalization, and patterns of soils are generally called soil associations. Usually the integrating factor of a pattern is some landscape process such as hillside erosion and deposition, or riverine aggradation and sorting. The landscape process is the integrating element which unites possibly diverse sites into a land system. The soil aspect of the land system is designated a soil complex or a soil association. In itself, a generalized pattern of soil provides only limited information on the details of local soil occurrences and provides a very limited predictability of soil data. However, if the user visits the scene and recognizes the pattern of the component sites of the land system, then he may assume for each kind of site the soil class that the pattern description predicts for it. Much of its potential predictive power then depends on the user's ability to distinguish sites within the pattern.

The procedure of choosing a soil classification that fits the landscape modes and boundaries seems desirable for many reasons. However, there is the risk of presuming the adequacy of the association between soil-class differentiae and the landscape-class differentiae by means of which they are to be identified during mapping. This association must be checked when creating the classification and later during mapping itself. To some extent the minimum spacing of soil observations during mapping may be adjusted to the degree of associ-

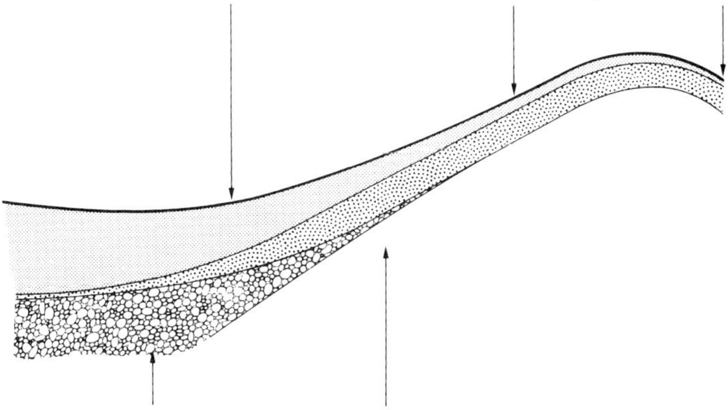

Fig. 2.10. Deciding on the soil boundaries to adopt in a soil survey is a real problem because the changes in soil profile are usually continuous over the landscape. The transect illustrates the possible dilemma in choosing between soil changes which match the landscape changes and those which are based on a 'general' or an 'exotic' rule.

ation between soil differences and external landscape differences; where the association is poor the graticule of soil inspection sites would need to be reduced to smaller intervals.

The second objection to taking the soil classification from the country is that it is imposed on the soil surveyor by the local interactions between soil and landscape and precludes him from using some pre-established and exotic soil-classification system for his survey. He may, for example, have been told or commissioned to use some established taxonomy because of its success elsewhere. But it is most unlikely that the soil modes and the limits of the sites he can observe in his area will correspond to the defined characteristics and keying values of the classes of the introduced system. This applies particularly to precise and hierarchical systems in which all the differentiae of higher categories also apply at lower categories. Presumably if such a system were applied many of the soil boundaries located would run across apparently uniform areas and be missing in some cases from apparent discontinuities. In effect, they would not fit the landscape. It is also likely that the association between useful and keying characteristics is less good than in the area where the classification was erected. The objection that a local classification imposes itself on the soil surveyor is well met by the greater objection that an exotic classification imposed on the project area will define areas which neither fit the landscape nor are likely to live up to their reputation gained elsewhere.

A third objection to taking soil classification from the country is that it will lead to problems of soil correlation, because each survey area is likely to throw up its own particular form of soil classes, and even 'similar' classes will differ somewhat between areas. But this is really an argument against correlation. Attempts to correlate slightly different classes from different project areas will lead to the same loss of precision as introducing an exotic key or class. We find then that the objections against taking the soil classification from the country are less serious than the alternative of conforming to an exotic classification, so we decide against the latter. The difficulties associated with the former can be met by accepting a taxonomic hiatus (see p. 40).

We have already emphasized that the grouping of similar soil entities into *soil classes* is different from the grouping of adjacent soil entities into *mapping units*. The concept of the 'polypedon' (Johnson 1963) attempts to bridge this gap between soil-profile entities and land-area entities. The polypedon is an area of land in which the component soil material exhibits a defined range of taxonomic characteristics. It is allowed to contain 15 per cent by area of deviant soils at series level or above, but apart from this its boundaries are where pedons of a different class of soil profiles are encountered. Thus the operative definitions of its boundaries are the same as the taxonomic differentiae of the adjoining soil classes, which is often equivalent to the imposition of exotic soil classes which are unlikely to fit the modes and boundaries of the actual scene.

The search for a merging or reconciliation between soil-taxonomic entities with landscape-area units has not been very successful; either we adopt a soil classification from the landscape itself, in which case it fits the landscape but not the rest of a soil-taxonomic system designed to cover a wider area, or we impose the classes of a general taxonomic system on the landscape, and then it is unlikely to fit the local soil modes or the individual character of the local landscape scene.

We prefer the former course, that is to develop and use the soil classification which arises from the landscape itself, because the classification will be used (whether for study, agricultural, or mechanical use) in the context of and in interaction with all its site characteristics.

However, we still see the need for an assessment of the extent of the association between soil and landscape classes.

To solve the problem of taxonomic correlation described above, we propose formal recognition of the fact that there always will be, and probably should be, a *taxonomic hiatus* between any general classification system and the local classes of any local soil-survey classification. It will often be convenient to indicate to which of the more comprehensive classes of the former each narrower local class most nearly belongs, but the differentiae of the latter should not be *altered* to coincide with the differentiae of the former. This hiatus will also apply between one soil-survey project and a neighbouring one, because even the 'same' class is likely to have some local peculiarities which would be lost if two or more local classes were jointly buried in a more comprehensive class, however it was derived from a comprehensive system.

High-level classifications

The creation of a national or regional classification system involves three essential operations:

(a) The development of a gallery of conceptual classes covering the full range of experience within a defined realm of soil experience;

(b) The rationalized elaboration of higher and lower categories of classes developed around the conceptual gallery. The categories of classes are formed by taxonomic chops (downward) and groupings (upward) of the original classes;

(c) The creation of a key to allocate unknown soil entities to their proper place in the system. In a few classification systems the key is the first and only stage in the development of the system.

The operations are listed in the order in which they are developed in most systems although there are exceptions. Some classifications consist only of (a), whereas others start with (a), then develop a hierarchy of rationalized classes as in (b), and then develop a key as in (c). At stage (b) some systems have only a few categories and rather general class definitions, while others have many categories and complex precise class definitions. The development of rationalized classes as in (b) is the development of hierarchies.

The development of keys as in (c) is usually the last stage, but a

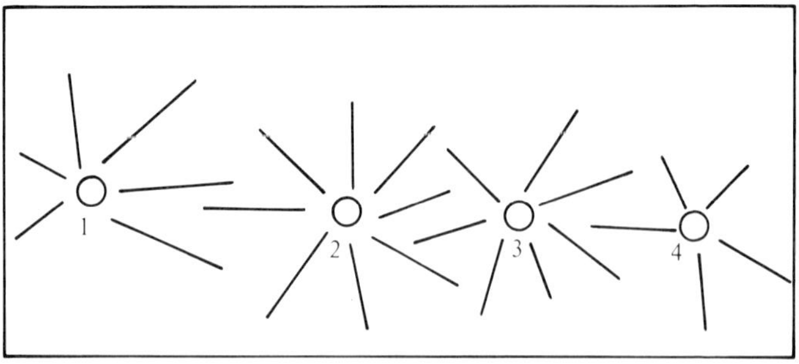

Fig. 2.11. A gallery of conceptual classes focuses attention on the nuclei of orderliness in a realm of experience and enlarges these as much as possible. Each class lacks boundaries. Parts of the realm of experience away from the central concepts can be assigned to the nearest class but the more marginal parts will have to be relegated to a miscellaneous or 'hold-all' class.

few systems start with their key, proceed to a hierarchy, and then generalize to classes after experience in the use of the key in a given realm of experience.

The creation of a classification system usually involves the inauguration of a matching system of names, either new words or old words that are given particular meanings. This seems an obvious need, and the more complex the system, the more complex the nomenclature is likely to be and the more necessary that it should have a clear rationale.

Gallery of conceptual classes

Generalized conceptual classes are formulated by the same sort of mental operations that most people exercise intuitively from a very early age to simplify their appraisal of, and response to, the multiplicity of data which presses upon them.

Initially at least the classes in the gallery of conceptual classes are felt rather than created logically; they are described rather than defined, and the perception of their central concepts is stronger than the realization of their boundaries. The classes are perceived as a synthesis of all their characteristics: morphology, soil genesis, environment, and soil process. The characters of these classes are thus determined by the realm of experience on which they are based and on the value

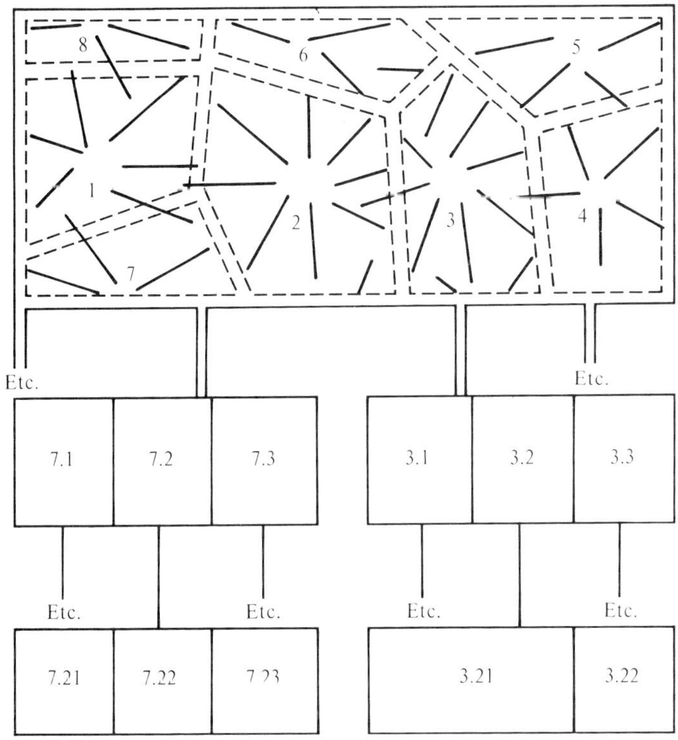

Fig. 2.12. In the rationalized elaboration of classes the conceptual classes are first converted into defined classes by defining the permitted variation of certain properties. They are then subdivided into lower levels of rationalized classes by subdivision of the range of other selected properties. This process of producing lower levels of rationalized classes of finer definition may continue through any number of stages.

judgements and predilections of the observer. This makes for full well rounded concepts, but the vagueness of their definitions makes it difficult to allocate new soil entities to their class unless they coincide with its central concept. Inevitably many soil entities are left in a miscellaneous unidentified category.

The rationalized elaboration of classes

Having created a gallery of conceptual classes, many classifications then develop a series of lower categories of classes by rationalized subdivision of the conceptual classes first perceived. The creation of new classes and new galleries of classes opens up new and almost unlimited prospects for elaboration. The operating principle is quite different from that used to develop the conceptual classes. In the rationalized divisions and subdivisions of classes, attention is given to the boundaries or demarcations between classes; i.e. to the choice of the point where the taxonomic 'chop' should occur and to the choice and definitions of differentiae and not to their central concept. Where there is gradation between one conceptual class and another the boundaries between them must now be defined.

Then these classes are subdivided to form a lower category of classes, each defined from its neighbour on differentiae that separate several themes of variation from a central 'ortho' concept. The subdivision of classes at one level to make a lower category of more narrowly defined classes may proceed through several stages, either introducing different criteria at different levels or using different degrees of particularity of the same criteria.

The definition of class boundaries and the selection of particular values of attributes to define the boundaries between the subdivisions of a class gives these properties and values priority over other properties and implies that these properties will also serve as indicators of the other properties of the class. This raises again the question of the orderliness of variation in the population. The positions of the taxonomic chops in regard to the orderliness that is present is all-important in determining the serviceability of the classes which they will produce. We have already shown that classes produced by subdivisions may be of little value if there was little orderliness in the soil population or if the chops were made in the wrong place.

The elaboration of a taxonomic system by means of selected differentiae and by an intricate system of precise definitions is attractive because it gives the appearance of precision and certainty, and the identification of a new soil entity is unambiguous. This certainty, however, is delusive. The prospects of predicting the useful attributes of the soil may be very good or very poor regardless of the precision of taxonomic differentiae.

Some soil taxonomists choose subclasses and the differentiae to separate them by reference to pedogenesis and soil processes assuming that, if the differentiae are indeed diagnostic of process, they impart usefulness to the classification. Thus *Soil taxonomy* (Soil Survey Staff 1975, p. 10) selects the differentiae of the upper categories among those properties which 'result from soil genesis'. Similarly de Sigmond (1938) and Muir (1969) are primarily concerned with producing subdivisions with pedogenetic relevance. On the other hand, genetic criteria are often considered to be too propositional and too often an inference rather than a direct measure of something; other authorities (e.g. Northcote 1971 and FitzPatrick 1971) then avoid using genetic criteria in soil classification on principle.

There are also disagreements about the extent to which rationalized classes should be developed into a hierarchy. Smith (1963) favours a hierarchy of elaborate construction; Macvicar (1969), Webster (1968), FitzPatrick (1971) and others consider elaborate hierarchies inappropriate to soil classification. Fig. 2.13 presents a general picture. Chapter 4 presents several existing systems. We have already shown above that, if classes are subdivided without appropriate analyses of data from the population, it is unlikely that the subclasses will facilitate predictions. The precision of subclass differentiae is not necessarily matched by any corresponding uniformity in the associated characteristics of the soil entities in each subclass. These may contain a lot of taxonomic noise and their keying properties may be the only ones their entities have in common.

Galleries of broad conceptual classes play a crucial role in the development of national classifications. A similar gallery of classes, narrow and more local in their scope, play a role of equal importance in each soil survey. At the commencement of each soil-survey project the soil surveyor examines the full range of soils in his area and formulates a gallery of soil classes which condenses his full experience in the most comprehensive way. This is also a gallery of conceptual classes. He will give much attention to checking the serviceability of these classes and their relationships to site features, land use, and lists of soil classes published elsewhere; he also examines the prospect of subdividing them into subclasses or of merging them into higher galleries of classes and then looks at the relationships of these to the national soil classification. Nevertheless he will be wise not to labour to close the taxonomic hiatus (Fig. 2.13). As indicated (p. 35) he will almost certainly lose information if he does. The information to which he gives access by correlating his soils via a common taxonomic grouping with the soils of other areas is often vague or rendered inapplicable through

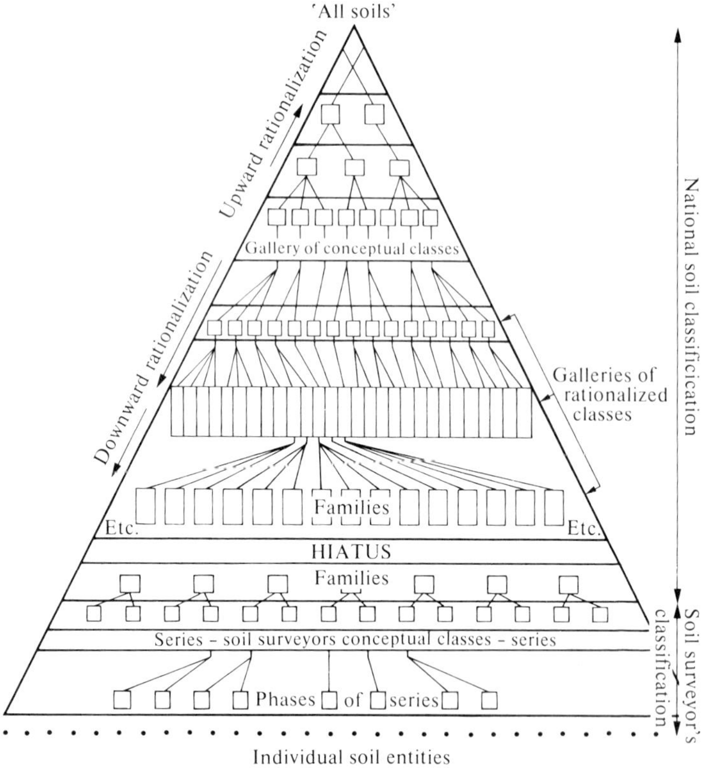

Fig. 2.13. A soil-classification hierarchy illustrating the form and some of the terms used in this context. As pictured the galleries of conceptual classes occur at two levels–that of the soil surveyor's series and that of a national taxonomist's great groups. Rationalized development of classes proceeds both upwards and downwards from these levels of origin. Upward development is by generalization and is of little consequence. Downward development is by taxonomic chops and makes most of the actual separations in a classification. There can be no confidence that the downward development from the national system will coincide with the upward development of the soil surveyor: hence the need for a formal hiatus.

their lack of correspondence in non-taxonomic features. The indications of failure of a classification system are when the useful properties usually associated with a class do not appear in new soils allocated to it.

It is important to realize that the soil surveyor's classes originate in his empirical concepts. He still needs to ensure that they are distinct from adjacent classes, to establish their pattern of soil-property relationships, and to test the nature of their role in solving the practical problem.

Proposals

Thus we have the major problem that it is deceptively easy to create an elegant soil classification to which any new soil entity can be allocated without ambiguity or hesitation. Yet this means only that every member of a class shares the keying attributes of that class. Unless all the entities within a class *also* share other attributes that are relevant to some land use, there will be no benefit in identifying them. This point has been made in several different ways in this chapter.

At the same time, unless the classes roughly correspond to the natural modalities of the survey area, and at least roughly conform to the more obvious discontinuities in the landscape, they may be difficult to map and require the subdivision of areas that common sense requires to be left undivided.

Undoubtedly there are some areas in which the soils lend themselves to useful classification and mapping. Equally there are some areas in which a classification or a map can only be meaningless. The problem is to find out which before carrying out a survey. Chapter 3 offers a procedure by which the surveyor samples his area to create a classification and legend and then samples it again to check that the classification and legend are meaningful. As proposed, the procedure will be applicable to many surveys at medium scale, but it is the principle that is important and not the precise details. Provided the surveyor accepts that he must establish that the local soil population contains enough orderliness to support a useful soil classification and provided that his check is unbiased, he may check it how he will.

3. Establishing the soil classification for a soil survey

Soil survey involves many different kinds of activity, but our main concern here is the process of establishing a soil classification for a particular local soil survey.

This chapter presents a number of major innovations introduced to rectify some of the defects of customary practice. They refer mainly to testing a first provisional classification in terms of its association with useful soil properties and the predictions that can be made from them; this ensures that the classes and mapping units of the survey that follows will be relevant to 'the problem' that led to its commissioning.

Simple outline

The following list of operations for developing a classification starts from the point where the soil survey of an area has been commissioned and its problem named. Certain complications and qualifications passed over in this brief statement will be explained in more detail later.

(a) A general review of the project area, its topography, geology, plant and crop ecology, and any prior or adjacent soil studies is carried out. Special attention should be given to 'the problem' and its relationship to any soil or other local factors.

(b) A provisional list of classes for the project area is generated. The soil surveyor examines the soil at numerous points of observed similarity and difference which have been chosen to cover the whole range of soils in the area. On the basis of this experience which includes consideration of site features, the incidence of 'the problem', and the soil profiles, he formulates the concepts of a set of classes to fit the area. The formulation of classes is an inductive, iterative operation.

(c) To check the provisional classification a sampling of the total soil-data pool (i.e. all the properties of all profiles) of the project area is carried out. The sampling must be unbiased and according to a fixed schedule of inspections, sampling, and laboratory analyses. The soil at each sample site is allocated to one of the classes already proposed. A miscellaneous class is adopted if needed.

(d) The proposed classes from (b) are tested against the sample drawn from the soil-data pool (c) to check that they are distinct and that they are associated with soil properties of interest and particularly with the soil factors associated with 'the problem'.

(e) The provisional classification for the area is recast and finalized in the light of the results of the testing programme. Some classes will be merged for lack of significant difference. If the enquiry so far has not established an association between the soil classes and 'the problem', further sampling may be undertaken. If this produces no evidence either, then the relevant information must be organized and presented by means other than classification. It will be apparent by this stage if a soil map will be useful or not, and the decision whether to map is now made.

(f) If mapping is to proceed, a soil key is prepared for the classification finally adopted and a map legend is prepared to link the classification to the landscape. Field technicians are familiarized with this key and they can proceed to map the whole area under the supervision of the soil scientist. The map should include a statement of its predictions and their reliability.

It is necessary to emphasize the distinction between operations (b) and (c). Operation (b) creates a tentative classification and operation (c) checks it against a fresh set of data. Operation (b) is intuitive and purposive; operation (c) is rigorously unbiased. Although they have the collection and sorting of soil and site data in common, they must be clearly seen as having different aims. Failure to keep them distinct may render the soil survey indeterminate.

The generation of the classes in (b) involves the examination of soil profiles at many sites and the tentative clustering of them so as to produce a grouping and ordering of soil variation which matches what occurs in the project area. Although the classification must cover the whole area, the soil surveyor need not necessarily examine the whole area in a regular way. He may develop and test hypotheses of soil variation and pay more intensive attention to certain areas, and to 'typical' soils and areas, than to others.

In contrast to this, the sampling of the soil-data pool in operation (c) must be probabilistic, evenly spread over the project area, devoid of any selection for 'typicalness', and free of any hypotheses about soil similarity and grouping. This allows the soil surveyor to collect a sample set of the soils of the project area which represents the whole area in its variation, its orderliness, and its attributes and their associations. This sample set will be used to test the tentative classification and associations claimed for the area by the soil surveyor.

Sampling for the generation of classes is called '*C sampling*' (C for classification); the data sets are the 'C sets' and the data go on a set of 'C cards' with their own C serial numbers.

The sampling of the data pool is distinguished as the '*P sampling*' (P for pool) to produce the 'P data sets' on the 'P cards' and numbered with P serial numbers. Table 3.1 tabulates the characteristics of each set and its use.

It should be clear that this principle—that a set of classes should be established on one set of data and checked on a second independent set of data—is absolute. The procedures proposed here, and particularly the procedures for checking the provisional classification, should be found useful, but they can easily be modified to suit particular circumstances provided only that the two sets of data and the two operations are not then confused.

Developing a provisional classification from the C sampling

At the beginning the soil surveyor should study all source maps available for his project area: geological, topographic, hydrological, ecological, agronomic, and land use. He should be looking for major and minor variations and patterns in the total landscape. He should look for the visible patterns of these differences and any new patterns that are visible in the aerial photos of the project area. The soil surveyor should give special attention to 'the problem' which motivated the client's request for a soil survey. He should study the relevant literature on the problem and examine the evidence of it in the field. As far as possible he should seek to identify any association of the problem with particular areas, kinds of landscape, or land-use records.

The soil surveyor will expect that the major differences of landscape and geology will be associated with major soil differences. He should plan his field traverses to cross these major differences, examining soil profiles, in order to ascertain what these soil contrasts are, and to acquire a broad perspective of the soils, and to formulate hypotheses about the soil profiles and their associations. The soil surveyor must relate the soils he recognizes to their expression as areas of land, with boundaries, for these are what will be mapped eventually. The largest area of land that can be *treated as if uniform* is often

TABLE 3.1.

Specification of the C- and P-sampling sets for soil survey

C-sampling set: the classification and hypotheses set.

Purpose: to create conceptual classes which initiate the classification scheme for the project; to generate hypotheses about the relationships between soil classes and soil factors; and to provide a background for the P sampling.

Procedure: to examine with soil auger and spade the full range of soil and site which the soil surveyor sees in his project area.
From this experience the soil surveyor develops concepts that group these soil entities into classes and he prepares a preliminary schedule of classes for the area. He checks these concepts by examining further sites, and he examines the soil variation within uniform-looking areas of each class and the nature of boundaries between his class areas.
On the basis of his provisional identification of the main soil attributes of interest in the area, and those associated with the problem, the surveyor prepares a scheme for the P-sampling.

Operational detail: Some 100 sites scattered fairly evenly over the project area are selected to illustrate typical and intergrade soil profiles and located on the map. The profiles are recorded from soil pits on C cards, each providing a full morphologic and site description. Data on within-site variations and the nature of soil boundaries are included where relevant. The soil entity on each C card is allocated to its class.

P-sampling set: the data pool and testing set.

Purpose: to create a sample of the soil mantle in the project area; to provide a soil-data pool from which representatives can be drawn to test the proposed classes and associations arising from the C sampling.

Procedure: to sample on a fixed grid of sites so as to produce a representative and proportionate sample; to collect a standard set of morphological, analytical, and site data from every site, the sampling and analytical scheme being based on the C-sampling experience.
Every sample site is located precisely by its grid intersection. Each site is to be allocated to a soil class of the C-sampling classification or identified as 'miscellaneous'.

Operational detail: The fixed grid is designed to give about 100 intersections or $20n$ sites, where n is the number of main classes in the C classification, evenly spread over the project area.
Each site is located on the map, allotted a P number, and all its data and analyses entered on a P card.

---

called a *site* (Clarke and Beckett 1971), and this is a convenient land unit to which the soil surveyor can tie his soil classification. He will wish not to subdivide a site in his mapping, and so the breadth of his lowest taxonomic unit often compares to the amount of soil variation which he finds within a site.

When the soil surveyor feels that he has acquired a comprehensive overall picture of the soil variation in his project area (which usually corresponds to the stage at which he begins to be able to predict the soil at a site *before* he augers or digs there), he goes into the field to 'typical' areas of his major kinds of country, locates himself centrally within a typical site, and there makes a full examination of the soil profile as revealed in a pit. He records this on a C card, describing also the site attributes and the soil variation he finds within the site area. This C record is allotted a serial C number and located on the map. This operation of selection, examination, and recording is repeated a number of times, both at sites judged to be duplicates of the first site (either near at hand or more distantly removed) and at sites perceived to be different.

In describing soil variability by forming groups at different categorical levels the soil surveyor must ensure that the range of difference in groups at each level conforms to established custom, as shown in Table 3.2.

Some soil differences occur in close propinquity, and others develop gradually over greater distances. Those that occur in close propinquity are often associated with the details of landscape relief—with land variations in slope, soil texture, and soil-drainage status (colour, gley

TABLE 3.2.

Ranges of soil-difference corresponding to category names

Amount of soil difference (least to most)	Category name (from lowest to highest)
Least informal difference	Characterized only by a symbol or inscription on the map
Greater informal difference	Phase
Least taxonomic difference	Soil series
Greater taxonomic difference etc.	Soil family etc.
High taxonomic difference	Great soil group

characteristics). They often form a characteristic pattern which is repeated with minor variations over large areas. By tradition slope and texture differences rate only a phase difference (a phase of a series) whereas colour and gley differences often rate a series difference.

The soil differences that develop gradually over greater distances are often expressed by changes in several profile attributes at once, and even by different forms of profile expression. They are associated with zonal or environmental changes in climate or ecology. Such profile differences are rated sufficiently important to warrant series or even higher ranking. From their nature it is difficult to draw a boundary between two series which differ in this manner, although the extremes of the range may be clearly distinguishable.

Clearly marked soil differences which are associated with marked geologic or topographic changes present no problem in classification or mapping. They are of different series, or even of different great soil group, even if the change and contrast is repeated frequently as a pattern.

The C sampling at the beginning is based on the first and highly tentative groupings and separations in the mind of the surveyor; his further experience during the sampling often leads to a consolidation and in some cases a confirmation of his ideas. It may also lead to a reshuffling of the cards, however, and to alternative groupings.

The surveyor also has the chance to confirm and check any propositions which he has developed about the association of soil attributes with site- or land-use elements. In particular, he investigates thoroughly any tentative associations that relate 'the problem' to soil, site, or management factors and he examines sites specifically chosen for this purpose. It would be reasonable to collect a few soil samples for laboratory analysis, although the main effort in this respect will be in the P sampling because the C samplings are too personally selected to justify a comprehensive laboratory schedule.

It would be suitable at this stage to allocate provisional names to the main soil series of the provisional classification.

Sampling the soil-data pool

The P sampling provides a basis for testing the classes and propositions developed by the soil surveyor in the preceding operation, the C sampling. Because it must be representative and relevant whilst being as economical of effort as possible, the whole project area is sampled

Fig. 3.1. Soil-profile and site descriptions together with selected analytical data are collected on a fixed grid of sites to build up the soil-data pool for a project area. The sampling must be objective and provide a complete coverage of the area.

with equal intensity and without any personal selection of sampling sites. These requirements are best met by sampling on the intersections of an equally spaced orthogonal grid. The origin of the grid within the project area should be located by taking random numbers, and the sample sites are then selected by strict measurement of distance along the axes of the base map or the air-photo sheets. Each site is allotted a P number, and these are recorded on the base map and the photo sheets.

The greater the number of P sites the better the sample, but laboratory analyses are involved for each sample and the number must be kept within the bounds of what is economic. One hundred sites or twenty times the number of main classes proposed by the C sampling, whichever is the greater, seems to be a reasonable compromise.

The list of soil and site attributes to be recorded and the samples to be analysed at each P site (the sampling scheme or A schedule) for each survey project will be decided from the soil variation revealed by the C sampling and the indications of association between soil and site attributes or between them and 'the problem'. The experience the soil surveyor has gained while developing his C sampling will help him to list those attributes which vary widely, those which may be associated with other attributes of interest, or those which are of consequence to 'the problem'. In general, the C sampling will have indicated the attributes to which detailed attention should be given.

The main soil attributes may be divided into

(a) those of use in allocating soil entities to their classes;
(b) those of intrinsic significance to the nature of these soils;
(c) those of significance to the incidence of the 'problem'.

This idea is developed further in Table 3.3.

The P-sampling schedule lists those attributes which seem to be most significant in the survey area and likely to give rise to significant or useful associations between 'identifying', 'intrinsic', 'use', and site attributes. Recording these attributes and collecting soil samples for analysis must be standardized for all sites. This requires that samples

TABLE 3.3.
Possible 'identifying', 'intrinsic', 'use', and site attributes of significance in a soil survey

Identifying attributes:	Those used in identifying the classes in the soil survey under execution. Probably include some of the morphological attributes of colour, texture, structure, consistence, thickness of horizons, concretionary material, or possibly site attributes.
Intrinsic attributes:	Those associated with many other attributes, such as the characteristics and amount of the exchange complex, the ions (cation and anions) associated therewith, the character and amount of organic matter, clay minerals, pH.
Use attributes:	Those associated with the 'problem' of this project, probably include some of available N, P, K etc.; wetness, hydraulic conductivity, available moisture capacity, crusting character, hardness, cemented layers, depth of soil, salinity, pH, triaxial shear strength etc.
Site attributes:	Physical attributes of the site. Probably include some of inclination, length of slope above site, texture of surface, stoniness, wetness (groundwater), vegetation both native and sown, provenance, geomorphic character etc.

should be collected at the same fixed depths at all sites. The experience of the C sampling is called upon to determine what these depths should be.

Once the sampling and description schedule is decided, the task of implementing it at the 100 or so sites is relatively routine and should be done as quickly as practicable. Trained field assistants can assist in this routine operation under supervision. It is important that the recorders make no attempt to select 'typical' sites. Wherever the grid intersection occurs, that is the sampling site.

It will be convenient to record the P data on cards, but the P-sampling schedules for different project areas and different 'problems' are likely to emphasize different soil and site attributes, so the design of the cards will vary too. The cards should have a central space for the formal description of the soil profile, the site, and the land use. They should also provide peripheral spaces for elaborations of selected attributes and the records of analytical data. There will be one P card for each site, and it will be convenient also to prepare summaries of certain kinds of data from many sites.

Testing the classes

Once the P sampling, the schedule of analyses, and the listing of the data on the P cards are complete, the classes developed by the C sampling can be tested. The classes and sample data are examined on the P-sample sets; the C-data sets are unsuitable because they are selective and because the classes are based on them. Each P site is allocated to its appropriate class in the C-sample schedule. Any sites that are truly miscellaneous are designated as such; the proportion of miscellaneous soils can be a significant element in a project area.

The tests and analyses we propose here involve only simple statistics. The reader is referred to Webster (1977) for more complex analyses. Only a few attributes will be examined for any one purpose, and it is important that the suitable and relevant ones are selected for each project area and its 'problem'.

The classes that the surveyor has created can be tested using

(a) Tests of the distinctiveness and uniformity of classes in terms of 'intrinsic', 'use', and 'identifying' attributes.

(b) Test of the association between pairs of attributes such as 'intrinsic' with 'use', and 'identifying' with 'use', and others, overall and within classes.

The first group of tests calls for histograms of soil attributes and comparisons of sample means. The second group of tests calls for scatter diagrams, and calculations of product-moment correlation coefficients for measured characters and contingency tables for ranked qualitative data.

Logically the first tests should be those that will show whether the classification has split the pool of data into groups which are internally uniform but different from each other, in terms of attributes which are of interest and concern to the soil surveyor. The choice of attributes should be made by the soil surveyor. He will select from among the 'use' attributes those which come closest to his understanding of 'the problem'. Having chosen the attributes to study, the class designation of each site in the P set is abstracted, and the corresponding values of the attribute are tabulated and examined. Dividing the range of values into five or seven groups such as 0–10, 10–20, 20–30, 30–40, 40–50 in Fig. 3.2 is usually satisfactory. Each value is allocated to its appropriate group, and the number of values in each group is plotted as a histogram or bar chart as in Fig. 3.2(a). The two peaks shown in the Figure suggest that there are only two distinct classes. A second set of bar charts is constructed, this time with the data for each class distinguished as in Fig. 3.2(b); this suggests that three classes are different in terms of the attribute chosen and the confidence limits of the population mean for each may be computed. The procedure for this is given in Table 3.4 and the calculation of the mean ($\bar{Y}$) and Standard Deviation (*SD*) are prerequisites. A two-tailed table of Student's t is also needed to calculate confidence limits. The procedure is repeated for a range of properties and attributes. The information enables the surveyor to decide which classes are significantly different, at what level of confidence, and in respect of which properties. In this way the provisional soil classification can be evaluated with reference to properties of interest. It may be found that some class distinctions are significant for one characteristic but not for another; if two classes do not differ with respect to any of the properties of interest, they may be merged without loss of information.

Such analysis tests the classification fairly comprehensively and judges the effectiveness of the classification with regard to 'the problem' and the potential use of the soil. We can also evaluate the classes in relation to the 'intrinsic' attributes of the soil. But it is to be noted that, if the analysis is applied to the 'identifying' attributes, it will provide no new information. However, it can be of interest in

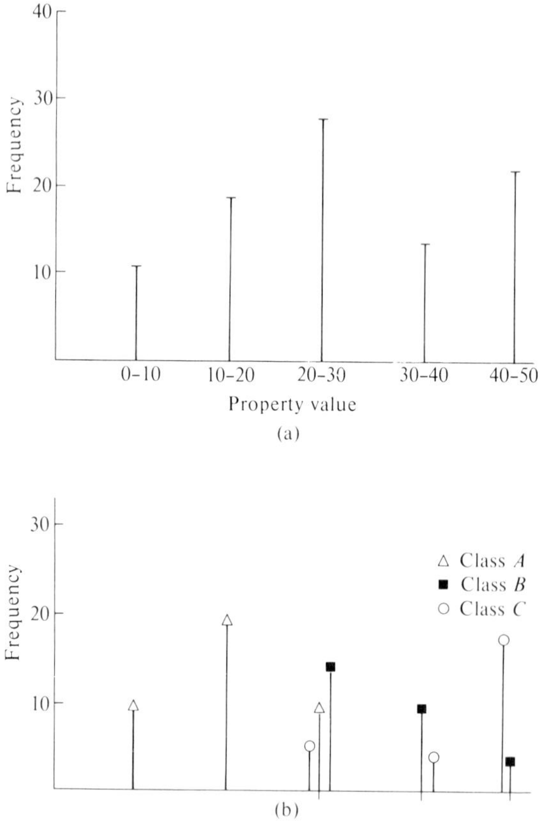

Fig. 3.2. The first test of the classes in a soil survey is by the frequency distribution of a chosen property of interest. If sorting of the data by classes (b) achieves a better segregation of values than is apparent in the population as a whole (a) the classification promises to be effective. Some of the pigeon holes would gather bimodal and tri-modal populations.

Fig. 3.3. The gallery of tentative classes which the soil surveyor develops during the C-sampling programme will need to be tested on class representatives drawn from the soil-data pool (the P-series samples). Tests of population and association statistics will be involved.

showing the degree of consistency that the soil surveyor has been able to sustain in identifying his soils or it can show him that many of his soil classes are not as highly characteristic as he thought but that they diverge from his ideals toward the miscellaneous or nondescript group.

The classes may also be compared in respect of their site attributes provided these are not used by the surveyor to identify the classes. Such an evaluation might enable the surveyor to assess the mappability of the boundaries on the soil map before the map is actually made. If the relationship between classes and site attributes is poor, the boundaries of the provisional mapping units may prove difficult to locate and need more investigation.

The foregoing procedures have met the philosophical problem of how to justify the subdivision of a continuous soil mantle. We have sampled the mantle and have only subdivided it on its own evidence that the null hypothesis (that there are no significant difference between classes) is untenable.

TABLE 3.4.

Calculations referred to in 'Testing the classes'

The mean (i.e. the arithmetic mean) $= \bar{Y} = \dfrac{\Sigma Y}{n}$

where Σ is the symbol for the sum of, $\bar{Y}$ is the property of interest, and n is the sample size.

Standard deviation $= SD = \left\{ \dfrac{1}{n-1} \Sigma (Y - \bar{Y})^2 \right\}^{1/2}$

Coefficient of variation $= \dfrac{SD}{\bar{Y}}$, usually expressed as a percentage, i.e.

$$100 \times \frac{SD}{\bar{Y}} \text{ per cent}$$

Confidence limits of population means

Upper limit $= \bar{Y} + X.SD/\sqrt{n}$

Lower limit $= \bar{Y} - X.SD/\sqrt{n}$

where X is a figure derived from a two-tailed t table entered for the desired confidence level and $(n - 1)$ degrees of freedom.

Product-moment correlation coefficient (r)

$$= \frac{\Sigma(Y_1 - \bar{Y}_1)(Y_2 - \bar{Y}_2)}{\{\Sigma(Y_1 - \bar{Y}_1)^2 \ \Sigma(Y_2 - \bar{Y}_2)^2\}^{1/2}}$$

where Y_1 and Y_2 are the two variables under study. Test significance of correlation coefficient r with tables for desired confidence level and $(n - 2)$ degrees of freedom.

Rank x column test of independence (adapted from Sokal and Rohlf 1969, p. 599).

For testing association of attribute Y_1 with attribute Y_2 where both properties are expressed in ranges, e.g. low, medium, high.

Prepare a contingency table of Y_1 against Y_2.

		Property Y_1 Low	Medium	High	Totals
Property Y_2	low	a	b	c	j
	medium	d	e	f	k
	high	g	h	i	l
	Totals	m	p	q	n = grand total

All cell frequences and totals are transformed to $\log_e$

1. Sum transforms of frequencies in the body of the table

$\Sigma_1 = (a \log a + b \log b + \ldots + h \log h + i \log i)$

2. Sum transforms of row totals

$\Sigma_2 = (j \log j + k \log k + l \log l)$

3. Sum transforms of column totals

$\Sigma_3 = (m \log m + p \log p + q \log q)$

4. Transform the grand total $= (n \log n)$

Then $G = 2$ {quantity 1 − quantity 2 − quantity 3 + quantity 4}

This value is to be compared with a χ^2 distribution with $(a - 1)(b - 1)$ degrees of freedom where a is the number of ranks and b is the number of columns. Enter chi square tables for degrees of freedom and significance level to compare with G for test of null hypothesis.

Association analysis

When there are differences in the behaviour of groups of soils there is likely to be some degree of association between certain identifying attributes and some of the intrinsic or use attributes. Pairs of identifying attributes may be associated also. It is desirable to find out more about these associations and the attributes involved because any predictions generated by the soil survey depend on them. Such information also leads to a fuller understanding of the different kinds of soil and their inter-relationships and provides clues for their better management and improvement.

In the analysis of associations of attributes among classes the soil surveyor will initially select those attributes that have already been found to be good discriminators. Using the terminology adopted above, Table 3.5 indicates some of the pairs of attributes which may be selected for analysis, and the kind of conclusions they would lead to if the value of the association is large and significant.

The soil surveyor would probably first examine the association between some clearly evident identifying attribute and land use, and the result of this would indicate the extent and confidence with which the classification could be used to make predictions about land use. Further analyses of this kind build up knowledge and justification of the classification being developed for the soil survey and test its strengths and weaknesses.

Association analyses are best preceded by biaxial plots (scatter diagrams) of the pairs of selected values drawn from the P cards. The soil class of each point should be shown since there may be dif-

TABLE 3.5.

Association analyses between pairs of soil attributes and the conclusions to which they may lead

Characteristics selected for study	Conclusions possible according to degree of association
Identifying × intrinsic	Pedogenetic processes
Identifying × use	Land-use prediction
Intrinsic × use	Ameliorative treatment and processes
Intrinsic × intrinsic etc.	Pedogenetic processes

ferent relationships for different classes. Fig. 2.4 illustrates the expected situation. Where the plot indicates association, as in Fig. 2.4(a), the data for each class should be abstracted and the product-moment correlation coefficient calculated (Table 3.4). The correlation coefficient *r* expresses the orderliness in the data within classes. However, it is only a single expression of association between two attributes, and a number of similar values for other pairs of attributes are needed to build up a general picture for each class. The surveyor will also be interested in the association between pairs of non-metric data. These can be analysed by contingency tables or the rank *x* column test of independence given also in Table 3.4.

Association analyses should be made on pairs of all the attributes of interest in the data pool because it is from a knowledge of such associations (i.e. of its orderliness) that the classification and map can be put to use. Such information can contribute substantially to the value of a soil survey, because the user needs to know what use he can make of the generalizations of the particular soil survey in its particular project area. If the soil classes have weakly associated attributes—a high taxonomic noise—little reliance can be placed on any generalizations concerning them. On the other hand, a validated group of strongly associated attributes means that a class can be used confidently for many generalizations and predictions.

As a result of these analyses the surveyor may find that he has too small a sample to decide whether certain pairs of attributes are associated. He may then make a subsidiary sampling (S set) for more data, provided he is fully aware of the risk of biasing his results by this sampling.

The results of some soil surveys are presented in the form of complexes of soil/landscape classes, these compound concepts being delineated on the map instead of individual soil-class areas. Such map areas represent a broader level of generalization than most mapping units on detailed soil maps. In principle, it is as necessary to justify the separation of the complexes as it is to justify the soil class areas; it is more difficult, however, because the complex is a group of diverse elements and the test of identity of *pattern* takes the place of the test of identity of a class of simple soils. A more extensive sampling programme would be needed to compare soil complexes.

Investigating 'the problem'

Many soil surveys are carried out to make land management easier or to say where management practices should be changed. We have already discussed the problem in broad terms. Here the narrower aspects related to soil-survey technique are considered.

There are two aspects to 'the problem' in soil survey. The first is to establish that the problem is associated with one or more particular soil variables in the project area; i.e. that the incidence of the problem matches soil variation. The second aspect is to find how it may be mapped in the context of soil classes and the other soil variables of the area.

The most common difficulty is the lack of sufficient data. There may be ample data on the intrinsic and identifying attributes of the soils of the area, but very little on the relationship between soil and productivity. Quantitative data on both productivity and soil attributes are needed to estimate the association between the two.

There are several ways of obtaining the use data. If the intended land use is practised in the project area, supernumerary sampling (S set) sites can be selected under this land use even though it occurs on only a selection of the whole range of soil classes. The soil surveyor should select some 30 or more sites to cover the full range of results from good to bad crop yield, or easy to difficult engineering problems etc., without restraint as to soil class or soil character. Each S site should be subject to the full P-sampling schedule, together with the measure or assessment of the productivity or problems at the site. The data should be recorded on S cards and then analysed for associations as above. Associations should be sought between land use and identifying or intrinsic attributes which, if they are found, become the basis for predicting land-use values and for designing the map legend and key. This assessment, however, is limited by the conditions of the sampling and studies of this kind can be very laborious; Howard and McIntyre (1946) is an example of such work. If there is no association between soil class and land use, productivity presumably depends on management and soil survey will not help solve any problems.

When there is no local evidence on the intended land use the soil survey of the project area must be supplemented by other data. There are several possible methods of collecting such data—to use evidence from other places where the soils are similar; to carry out field experiments in the area; or to deduce the land-use potential of the soils from their attributes. However, there are problems. Associations between soil attributes and land-use values found elsewhere may be unreliable in a new area, and the associated properties of similar looking

Fig. 3.4. Differences in crop production may be the reason for undertaking the soil survey, but if the grid sampling does not cover the range of crop responses with enough data for a proper analysis, a supernumerary sampling (S series) is needed. Sites should be chosen according to their high and low production and not according to their soil. Analysis of the data should seek association between crop quality and soil factors or soil classes.

soil from different project areas may be quite different. An exotic diagnostic may be a valuable guide to experimental work, but it is no less onerous to confirm its applicability to a new area than to establish it *de novo*.

The direct determination of the performance of the soil in its intended use is undoubtedly the most reliable way of obtaining the desired information, but it is laborious and time-consuming to establish field plots of a standard treatment involving the 'problem'; e.g. growing the crop over the range of soil variation which the soil survey has revealed. The yield or cost data obtained can be tested for their association with the soil classes and soil attributes as above. Studies by Loveday and MacIntyre (1966) and Colwell and Esdaile (1968) are of this kind. Establishing reliable diagnostic associations with land use is expensive, and it has been done for only a few soil surveys.

The third approach—the *deduction* of soil potential from commonly known general relationships between productivity and soil attributes—is often adopted to provide predictions. This is really a layman's solution to the problem and exercises none of the professional standards that have been maintained up till now. It may succeed, however, if the soil morphological attributes are the main soil factors in 'the problem'.

In extreme cases, soils can be judged unsuitable for use because of stoniness, salinity, or waterlogging etc., and the finesse and powers of discrimination of the soil-survey technique are not called for to determine this. In contrast, several soils may have no deducible limitations to use for the purpose in view, and the surveyor would readily give them a high rating. It is between the sector of obviously good and obviously bad that a soil survey has its special role and where its success can be assessed. Providing the soil variability of the project area is orderly and provided the surveyor has access to sufficient soil/use relationship data, soil-survey technique allows a finely graded evaluation and a precise prediction of soil behaviour in their environment. A deficiency of data on the soil/use relationship often limits the effectiveness of soil survey. Some soil surveyors try to compensate for this deficiency by either calling for more analytical data on their soils or conducting auxiliary glasshouse or field trials, but these may miss the primary need for sets of yield data and associated soil data from the full range of the soil variation in the project area. A choice of one or two 'typical' sites alone for field trials is not enough to give a confident result.

Many soil surveys fall short of establishing the associations between soil classes and use although this should be their main objective. No amount of discussion of local pedogenesis will compensate for this deficiency.

Finalizing the classification

When the surveyor's original gallery of conceptual classes has been checked and he has completed his detailed assessments of the significant soil groups and the association of 'the problem' with the soil variables, he can decide what changes he should make to his provisional classification. Some classes may be merged with others because they cannot be distinguished using the sample data; others may be merged because the difference, though present, is not relevant to the purpose of the survey.

The scatter diagram of attributes in some classes may suggest that a better grouping of the soil could be made than that intuitively adopted. For instance, plots of an identifying characteristic and an intrinsic one may prove to be similar to the situation shown in Fig. 3.4, and it may be possible to make a more appropriate chop than that previously adopted.

Fig. 3.5. The soil surveyor will make the final adjustments to his classification with the complete record of data and the results of the tests of his classes before him. Some of his classes will be merged because their separation cannot be justified, and some perhaps will be subdivided for practical reasons.

The surveyor may choose to subdivide other classes because of information such as slope or depth which is peripheral to the soil but central to land use.

The main soil classes that remain should be formally recognized as soil series and given a locality name, e.g. Smithtown series. Subdivisions of original classes on texture, slope, stoniness, or depth should be recognized as *phases* of their original series, e.g. the sandy loam phase of Smithtown series. Those classes which are clearly groupings of two or more original classes grouped on their similarity may be designated as soil families and named after the most characteristic of its series, e.g. Smithtown family. Where, for convenient mapping, original classes are grouped together in spite of their differences because they consistently occur in propinquity, they may be mapped as a complex or association, e.g. Smithtown–Clare complex.

The final contribution to the classification of the soils of the project area is to match its soil series and families to the series of any adjacent survey area and to the appropriate soil classes of any regional or national soil classification if such exists. Exact identity of soil classes is not to be expected, as explained above, nor could it be presumed that classes that seem alike on their identifying attributes are in fact so in all their attributes. An accumulation of soil surveys in a continent will lead to an ever increasing list of soil series, and eventually it will be necessary to marshal these into a system to make the best use of the information they contain. This could be done by grouping the series into higher classes (possibly at more than one level) on the basis of common attributes, probably of characteristic horizons as in the South African system (p. 120). This would leave the conceptual origin of the system at the series level and not at the higher great soil-group level where many classification systems place it. The advantage of the former is that the validity of the series can be tested, in the latter it cannot.

As amended the classification should be prepared as a comprehensive statement about the soils of the project area for presentation in the memoir of the soil survey. The statement should list and describe the series, firstly on their central concept and then if possible on the limiting values of the identifying attributes determined by the soil surveyor's original examination of the classes and study of the C and P set card decks of the classes. The statement must list all his estimates of means and dispersions, and any associations of properties that the study of the data pool has revealed. Special attention should be given to the presentation of the findings regarding 'the problem', its association with the soil series, and with their identifying and intrinsic properties. These data must be presented with their corresponding confidence limits.

Preparing the soil key for routine mapping

With the classification finalized it is time to make the necessary decisions about mapping. Mapping is the customary way of describing and displaying soil variations. If the statistical analyses outlined above showed little contrast between the classes or much variation within classes, then presenting them as separate and ostensibly distinct units on a map may not be a useful way of communicating information about them. Classes that are especially variable may have to be designated 'miscellaneous soils' and merged with others of like designation. Decisions on whether to map and what series to map will also be based on how successful the surveyor has been in relating 'the problem' to mapable soil classes or variables.

Mapping should be done in a consistent way once the basis has been established. It would be quite undesirable for the soil surveyor

Fig. 3.6. Given a classification and a key the soil technician is in a position to identify the soil at any point on the minimum number of properties.

to make new 'interpretations' or combinations of soil attributes as the mapping proceeds. The soil classes established from the sample must be identified consistently. This is a routine operation that can be done by technical officers with the soil scientist acting as supervisor. A soil key enables or assists the technical officer to allocate any soil profile to its correct or most appropriate class, series, phase, or map unit as finally adopted for the soil survey. To produce the key the surveyor must identify those attributes by which each class can be distinguished from all others. The C and P cards are sorted into decks each representing one class as finally accepted. He then chooses the two classes most alike and sorts through the card deck of each to identify the one or two characteristics which most consistently distinguish them. This process is repeated for other pairs of similar classes until all classes, including the most dissimilar ones, are allotted distinguishing characteristics. A key is then devised which will, by a chosen sequence of discriminations, separate each class from the others. The classes of peculiar character are separated first by one of their special features, and then the more similar classes are separated on the remaining differentiae.

In devising a key it is reasonable to adopt, as far as possible, the most accessible differentiae which occur in the upper part of the soil profile or are visible without digging a hole. This makes for economy of effort.

The technical officer covers the area on 'within-sight' free traverses, preferably with the aid of aerial photographs, and visits in turn every area of apparent uniformity. Using a soil auger and the criteria of the key, he identifies the soil at a site in each subarea. Apart from identifying the soil classes at subsites, the compilation of the map requires boundaries to be drawn. The nature of the boundaries will have been investigated during the sampling for the C set. By referring to the data for each final class, the surveyor should be able to judge which of the boundaries apparent in the field are compatible with the classes to be identified with the soil key. If the boundaries are likely to be diffuse or if the associations between soil and site attributes are weak, then more inspection sites must be checked than those where the boundaries are sharp and the associations strong. A convenient field working scale is 1 : 20 000.

The use of a key may be expected to cause some loss of predictive power. An overall assessment of this could be made by testing the field technician's capacity to identify the classes of all the cards in the P deck, with their class names covered, using the key. This test would of course be made after the technician had been trained to the project standards. The result would perhaps best be expressed as a percentage of correct identifications for the whole array of classes. The possibility of personal error in the use of the key would need to be investigated.

The legend

The legend is the explanatory tabulation which goes on the soil map to describe the soil-mapping units and to identify any mapping symbols and inscriptions. The legend need not offer a full statement of the soil classification, but it must indicate which soil classes are to be expected in each mapping unit. It should also list the soil content of all undifferentiated soil complexes and miscellaneous soil areas as fully as the available information allows. The legend will refer mainly to the soil series and phases that are mapped. Any soil families that are mapped would also be included in the legend.

The soil key should be available in the legend or memoir so that the user of the map can check the soil class at sites of interest to him. It is desirable too that, for each series mapping unit listed on the legend, there should be some assessment of its possible error such as that which is indicated by its association analyses and the error involved in keying.

Appraising the error in the soil-survey project

The soil map and soil classification act as guides to the available information about the soils of an area. Whether his interest is intellectual or practical, the user will want to know to what extent the information set before him is accurate. Some statement of reliability should be given about the published map and the accompanying memoir. Unfortunately this is not standard practice. On the contrary, the user is often led to believe that all the information is absolute, all associations perfect, and that the soil boundaries and soil classes are real. It might appear that all the soil variation is orderly, and that the soil survey has found the orderliness in the soil variation. Soil-classification systems, especially those working on a keying principle, have no place for nondescript soils, and they offer no way of assessing the miscellaneous component in soil-data pools. In such a context, the suggestion that there may be miscellaneous and nondescript entities and sites is not acceptable, and their appearance is put down to a faulty classification system or to faulty workmanship.

But the surveyor or user cannot assume a high degree of order in soil-data pools. As cited above, Webster and Butler (1976) found that only about 20 per cent of the soil variation at Ginninderra could be accounted for by soil classification. In that case a soil map of the area could at best display something less than 20 per cent of the variation; mapping the soils at Ginninderra using conventional practice would not be worthwhile—a conclusion also reached by the authors.

In the United States soil-survey procedure allows an error or uncertainty factor of up to 15 per cent in mapping. However, no formal procedure for calculating the error is presented, so we know very little about the actual errors. Wilding *et al.* (1965) found 60 per cent error in the mapping of similar soil types in Iowa, and an average error for all delineated areas was 30 per cent. However, simple errors of identification or mapping are not the only nor perhaps even the main errors in soil survey. Using 'error' to mean the proportion of uncertainty or unpredictability in a classification or map, we can identify several components in the total error:

(a) *Uncoordinated variations in the soil-data pool.* This variation is a fact and however we view it it cannot be altered.
(b) *Error created by classification.* We have seen above that taxonomic chops, if misplaced, increase disorder.
(c) *Error created by the act of mapping.* Errors in translating

the classification into identifications and boundaries on the map.

The surveyor may find it difficult to assess his errors in these terms, but he has some means of assessment within his field of action, and these he should use as an indication of error:

(i) a statement of the proportion of sites in the full P set of cards that fall into the miscellaneous class;

(ii) the additional proportion of the P set of cards which the field technician misidentified, or wrongly identified as miscellaneous, when using the soil key with unlabelled cards;

(iii) the mean correlation coefficient for the associations of pairs of attributes of interest within each of the major soil classes of the project area. The pairs of attributes should include a number of 'identifying', 'intrinsic', and 'use' pairs.

Illustration of subroutines in the establishment of a soil classification for soil survey

In this section we illustrate some of the recommended subroutines which we have already presented in general terms. The illustrations come from the soil survey of County Moira in Victoria (Butler *et al.* 1942). Not all the subroutines were carried through as a historic fact, but the material and relationships from that survey can be used with advantage as illustrations. We use the same terminology as that adopted in the preceding sections.

The C sampling

The soil surveyor traversed the project area broadly to familiarize himself with the soils and to enable him to draw up a series of soil classes which covered the range and modes of variation occurring there.

The soils in County Moira are broadly similar; they have a clay *B* horizon and lime (as amorphous and concretionary material) occurs below a depth of 70 cm. The variations that occur are in texture, structure, colour, consistency, and in the thickness of horizons. The *Moira series*, the most widespread soil, occurs in a characteristic landscape of featureless plains. It is a brown loam, rather dense, passing abruptly at 14–18 cm to a reddish brown clay which is firm, dense, and plastic when moist. This material continues with little change to a depth of 1 m. Small pockets of lime occur from about 70 cm.

Most of the other soils can be described by their differences from the Moira series. Moira soils vary from site to site, of course, and when the variation becomes too great the surveyor judges that a different class must be recognized. Such a difference occurs with soils of sandier texture. The *Cobram series* is recognized by its fine sandy clay loam texture below 75 cm, its slightly more friable clay in the *B* horizon, and its rather deeper and sandier *A* horizon. Cobram series is also distinguishable from Moira series from its landscape, natural vegetation, and agronomic responses. It occurs on slightly elevated sinuous ridges in contrast with the plain of the Moira series.

The tendency to more sandy soils is carried further in some sites adjacent to the Cobram soils and reaches the extreme where almost the whole profile consists of brown sand with a weakly developed clay horizon at about 50 cm. This is the *Sandmount series* which is associated with sand dunes and a contrasting natural vegetation.

This spectrum of texture from the loams and clay of the Moira series to the sands of the Sandmount series is subdivided at a number of intergrade divisions. Some intergrades are given the status of types, some are phases, others are recorded only by map inscriptions according to the amount of variation involved (see p. 46).

The plains of Moira soils have, as contrasting sites *within them*, small slightly depressed areas. These are wet for short periods after rain, and the vegetation contrasts with that on the plain. The soils here are generally similar to the Moira soils, but their colour is grey brown and yellow grey mottled instead of brown. They are dense, very hard when dry and firm, plastic when moist, and they tend to gilgai microrelief. They are named *Naringaningalook* and *Boosey series*, respectively, the latter being greyer and more of a contrast from the Moira. This trend to wet poorly drained sites is carried a step further in the *Muckatah series*, which is a grey, mottled, and coarsely structured clay with gilgai microrelief.

There is another difference from the Moira series in a broad sweep of country in the south of the project area. The landscape and native vegetation show some insidious contrasts, but the differences are not incisive. The attribute of clear contrast is the specially high degree of friability of the subsoil clay. This material has the handling properties of a loam, but it is not sandy like the Cobram series. The friable attribute extends upward from the subsoil and, where most fully expressed in the *Katamatite series*, the *B*1 horizon as well as the *B*2 is

Muckatah

Moira

Ulupna

Cobram

Yarroweyah

Sandmount

Plates 1 and 2. These illustrate the landscape sites associated with the named soil classes of County Moira. The difference between sites is often subtle and is not merely a matter of vegetation (though this is involved too). The evenness of the surface, its shape, texture, and colour are all to be taken into account.

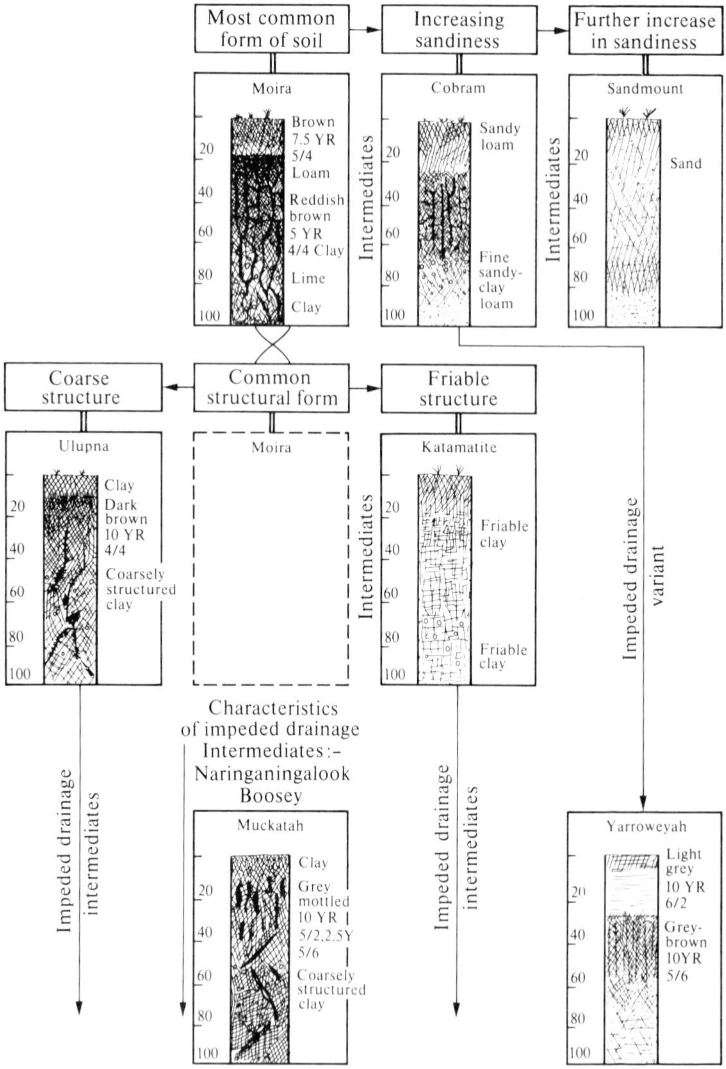

Plate 3. The morphological relationships of the main soil classes of County Moira. The Moira series may be taken as the reference point of soil variation, and each soil class is described on the attributes where it differs from Moira. The main themes of variation are texture, drainage, and structural expression.

affected. Where only the lower part of profile is affected, the situation is met by establishing a 'friable subsoil phase' to Moira loam. The friable subsoil seems to result from a parent material difference because other soils in the same locality—Naringaningalook and Boosey series—also have friable subsoil.

Another variant from the Moira class occurs as a broad sector along the northern side of the surveyed area. Here the landscape is mainly a plain, but differs from the Moira type by the presence of flood-retreat gutters (a history of seasonal overflooding from the Murray river) and a contrasting native vegetation. The soil profile has a number of minor contrasts from the Moira: it is dark brown rather than brown, the *A* horizon is a light clay only about 10 cm thick, and the whole profile is of a dense coarsely structured clay. It is called the *Ulupna* series. Associated with the Ulupna series, on slightly lower sites but still on the plain, is a dark grey gilgai clay soil called the *Mywee* series. In the Ulupna-Mywee landscape the occluded drainage areas have a soil similar to the Muckatah series as it is mapped elsewhere in the project area.

Landscape structures like the Cobram sites also occur in this flooded section of the area, but the soils lack the brown and red brown colours of the Cobram series and are instead light grey in the *A* horizon and yellow brown in the *B* horizon. They are named *Yarroweyah* series.

There are a number of minor named and un-named groups of soils in the area which hardly warrant a mention here.

Plates 1-3 illustrate the results of the C sampling of the soil survey of County Moira (Butler *et al.* 1942). Plates 1 and 2 illustrate the landscape sites associated with six of the soil series, and Plate 3 illustrates seven of the soil profiles and their relationships to each other.

The County Moira soil survey was intended to provide technical assistance in planning the area for irrigation development to horticulture and pastures. Since the irrigation planners did not specify their soil requirements, the soil surveyors had to develop their own criteria for assessments. Fortunately a small area had already been irrigated to horticulture for over 15 years. From the size of the fruit trees and their reputed yields, it was apparent that the texture of the soils determined their productivity. Under irrigation the more sandy soils were most productive and the more clayey soils least productive.

This observed relationship between irrigation success and soil texture confirmed the use of soil texture as a major discriminant in the soil classification, and it even prompted further phase divisions and finer subdivision by map inscription. Identification of poorly drained soils was regarded as being of self-evident importance in irrigation predictions. Unfortunately none of the irrigation occurred on the types of soil with friable subsoils so there was no indication of the role of that characteristic in irrigation.

The P sampling

This sampling (to establish a sample of the soil-data pool) was not made in the County Moira soil survey, but we can consider the issues that would have been involved in such a sampling using its materials and circumstances as an example.

The total area of the survey is 75 000 hectares, and a grid of about 100 sites on this would involve a grid spacing of 2.8 km. The procedure for making an unselected even distribution of sites over the project area need not be enlarged on; the graticule of latitude and longitude could be used.

The main decisions of the P sampling are in the schedule of attributes to be described, sampled, and analysed at every graticule intersection. After making a standard description of the profile and site, profile features of particular interest are chosen for special examination and finally horizons of interest are selected for measurement and for the collation of soil samples for laboratory analysis. All of these selections are made in response to the kinds of soils, their variation, and the information needs of the 'problem' in the particular project. In the County Moira project the morphologic attributes we would select for special interest are texture and structure of the *A* and *C* horizons and the colour of the *A* and *B*2 horizons.

Laboratory analyses are necessarily restricted because of their cost, and in addition the sampling is restrained by the need to have compatible data for statistical analyses. In the County Moira survey we would select sampling depths of 0–10 cm for the *A* horizon, 25–35 cm for the *B*2 horizon and 90–100 cm for the *C* horizon. The selected analyses would be those likely to be significant to irrigation (e.g. hydraulic conductivity, clay content, and salt) or likely to be intrinsic to the general pattern of the soils in the project area (e.g. the exchange characteristics, cations, pH, and clay content). The proposed *A* schedule for the P sampling and analyses is given in Table 3.6. At each site five samples would be taken for analyses—two of them undisturbed core

TABLE 3.6.

Notional A schedule of description, sampling, and analyses for the P sampling of County Moira soil survey

1. Standard description of soil profile to 1 m depth or more as appropriate and of site.

2. Standardized ranking of profile features:

Depth (cm)	*A* horizon *C* horizon	Texture *A* horizon *C* horizon	Structure *A* horizon *C* horizon
Colour	*A* horizon *B*2 horizon	Site character	

3. Sampling (1) 0–10 cm for *A* horizon
(2) 25–35 cm for *B* horizon
(3) 90–100 cm for *C* horizon
(4) Undisturbed cores 10 × 10 cm from 0–10 cm and 25–35 cm
} from each site

4. Analyses:

	A	*B*	*C*
Clay content	√	√	√
Hydraulic conductivity	√	√	
Salt (electrical conductivity)pH	√	√	√
Exchange capacity	√	√	
Exchangeable cations, Na, Ca, Mg	√	√	

samples for hydraulic conductivity and the other three bulk samples for the other analyses listed.

Testing the classes

This was not done in the actual County Moira soil survey, but it would be of interest to demonstrate what the tests would have been and might have revealed according to later research in the County Moira and in adjacent areas. The first tests on the P sets would be the frequency distribution of hydraulic conductivity in the whole set of samples for sample *A* and for sample *B*. Each point would be identified by its series to allow the separateness and significance of the soil series to be assessed with reference to these important use attributes. A similar test would be made for salt in sample *C*.

Frequency-distribution analyses of the data for exchange capacity and exchangeable cations for both *A* and *B* should indicate the effectiveness and consistence of the classification in relation to 'intrinsic' attributes. Tests of association would be made on the texture of sample *A* paired with the hydraulic conductivity of sample *A*, the same pair of attributes for sample *B*, and the texture of sample *C* and its salinity. Tests would also be made on the hydraulic conductivity of sample *B* paired with exchangeable cations of sample *B* and clay content of sample *B* paired with exchangeable cations of sample *B*.

Investigating 'the problem' in the County Moira soil survey was not a sophisticated operation, but it would be interesting to use the material of that survey as an example of how the problem may be met.

'The problem' was to assess the different soil classes in their suitability for irrigation. There was no general practice of irrigation throughout the area, so there was no prospect of using the survey technique to establish associations between soil class and irrigation success. The area of some 70 hectares of established irrigation orchards occurred in one compact area in the north east of the project area, and here the surveyors *observed* a relationship between soils and irrigation success. This observation could be expressed in terms of sandy textures being favourable and dense clays inimical to high productivity; or it could be stated as Sandmount and Cobram series being good for peaches, but Moira series being barely suitable and Boosey series quite unsuitable. These assessments are obviously of limited status–the sample base is clearly inadequate. Indeed, soils of the friable subsoil character were not represented at all, and the Ulupna series was represented by half a hectare. The assessments of all the soils were nevertheless based on these observations, together with deduced associations between soil and site attributes and irrigation suitability.

The irrigation development that followed took place many years ago and the result has generally been satisfactory. However, there have been crop failures in large areas where the district groundwater levels have risen to cause waterlogging and loss of production. Groundwater levels were not studied at the time of the soil survey and indeed they were not within the range of the augers used.

In the same broad region as County Moira, Howard and MacIntyre (1943) made a statistical study of selected horticultural soils at Griffith and found texture–the main soil variable–related to tree growth. The coarser textures were associated with the greater growth. However, Loveday (1964) worked on a narrower range of more clayey soils at Deniliquin and found a *positive* relationship between yield (of lucerne) and clay content because of a high correlation between other soil attributes and yield, including the 'use' attribute of incremental available moisture.

Loveday and McIntyre (1966) working with clover on 23 sites with a selected range of soils within a radius of 8 km near Griffith found that water entry limited yield. They also found an association between soil texture and yield; the relationships with fine sand and silt and also with salt content and exchangeable sodium were all negative.

This variation in the form of the relationships between yield and soil texture, even within the same broad region, may be due to different associated attributes and to the different ranges of one attribute prevailing in the different localities. It emphasizes that, if soil classes or soil maps are to have any predictive value, they must be based on real criteria and checked locally, and that it may be very unwise to extrapolate any relationship from one region to another.

4. Some established soil-classification systems

The soil-classification systems which are reviewed here are all *national* classifications, i.e. they refer to the range of soil variation occurring over a national or continental area and they operate at a high level of generalization. They marshal generalized soil concepts into a gallery of classes which is one logical step downward from the category 'all national soils'. From this gallery of classes they extend down to levels of increasing particularity, and some of them reach the categorical level at which the soil surveyor generates his gallery of local soil classes. All classifications have certain taxonomic structures:

(a) A gallery of conceptual classes—the starting point of most classification systems;

(b) The rationalized elaboration of classes; this usually follows (a) and leads to a more or less comprehensive hierarchical structure;

(c) The development of keys to allocate soil entities to classes. In some systems the key *creates* the system of classes.

In some cases the zone between a soil surveyor's series and national orders or great groups is scantily covered by a few loosely defined categories of classes. In others there is a complex multiple system of categories of classes—a big hierarchy. Many of the major differences between soil-classification systems correspond to different stages in the history of the subject. The development of ideas has been very uneven, and some current classifications are still based on earlier ideas, while other systems present later innovations. Classification systems also differ widely in the weight they give to *idealism* or *empiricism*. The idealist states that a basic theory of soils is the proper foundation for classification. An empiricist takes soil groupings as they seem to occur in the field without fitting them to a soil theory. De Sigmond (1938) and Muir (1969) illustrate the idealist stand, and the systems of FitzPatrick (1971) and Northcote (1971) illustrate the empirical stand.

Historic trends

The 1950s were times of dissatisfaction with soil classification. Leeper

(1954) complained of the ambiguity of the systems in use at that time; this he attributed to ambivalent class criteria which then ranged between genetic process, morphology, and environment, and lacked precise definition. Similarly the United States soil survey began to develop a new classification system in 1951 (U.S.D.A. 1975, see Foreword) because the existing system was 'inadequate in precision to the demands made upon it' and the criteria for families and the higher categories were poorly defined. Smith (1963) strongly advocated a well defined hierarchical system to extend from the polypedon upwards to soil orders.

These criticisms were provoked by classification systems like Stephen's (1953) *Manual of Australian soils* or that of Baldwin, Kellogg and Thorpe (1938). Both these systems consisted of galleries of conceptual classes defined only by their central concepts; this led to the lack of precision in definition and allocation inherent to such systems. Such systems are a natural first response to the need for a classification and they have the advantage that each class is a well rounded concept with all attributes included. But it is hard to define their classes and a new profile may appear to match the descriptions of more than one class. It is not altogether surprising then that those who were looking for orderly administration in a soil-survey organization of continental dimensions were dissatisfied.

The American reaction was to create a comprehensive rationalized hierarchy of classes in a series of publications leading up to *Soil taxonomy* (U.S.D.A. 1975). This has six categorical levels of fully defined classes and keys to identify all classes. Northcote (1960, 1971) took up some of Leeper's suggestions. To overcome the uncertainties of a system based on a gallery of conceptual classes, he started instead by promulgating a key for a system which covered five categorical levels and provided pigeon holes for 855 classes in the lowest categorical level. These classes were closely defined on relatively few differentiae but not described. It was expected that class descriptions would build up as a result of experience with the key.

Both *Soil taxonomy* and Northcote's system aim primarily to avoid uncertainties in the allocation of a soil entity to its class. At all costs they seek to avoid ambiguous and miscellaneous allocations, and their classes are created by a system of class definitions and taxonomic chops. In this kind of system there is a risk of excessive proliferation of classes with high taxonomic noise (pp. 23 and 39) and the value of the classes depends on significant associations between soil attributes in the soil population (pp. 5 and 28).

Jones (1959), Webster (1968), Knox (1965), and FitzPatrick (1971) have criticized *Soil taxonomy* and its style of classification. Their main points of criticism are on the strongly hierarchical nature of the system, the narrowly defined classes, and the lack of accommodation for intergrade soils, which they consider to be incompatible with the nature of soil. In some cases the criticism is directed at the attempt to reconcile the concept of 'polypedons' with the lack of any soil entity in nature.

FitzPatrick (1971) goes furthest in developing an alternative system that makes a minimal presumption of orderliness in soil data. He does not presume even the unity of the soil profile, but accepts as a 'pedounit' any combination of horizons which is found to occur frequently in an area. If similar pedounits are encountered frequently during the orderly examination of an area, the group would be acceptable as an *ad hoc* soil class (a series). Similarly, *ad hoc* groupings of these at a higher national level may give rise to world classes.

Our reaction to the excessive development of hierarchies is similar to FitzPatrick's. The search for prescision in allocating new entities to classes and hierarchies is misspent, and we would prefer to search for systems which make minimal presumptions of order in soils and which accept empirically the entities and classes which experience throws up.

We believe that it is unrealistic to presume a high degree of orderliness; indeed this assumption can easily lead to the multiplication of soil classes that lack any of the predictive attributes which we expect of them. The creation of soil classes, in both soil survey and soil classification, carries with it the obligation to prove their reality by appropriate tests on samples representing the area in which the classes occur. The remainder of this chapter is applied to the description of a range of classifications, and comments on the extent to which they seem vulnerable to the criticisms of Chapter 2.

The United States 'Soil taxonomy' (U.S.D.A. 1975)

This comprehensive system is the culmination of many years of development and has some pretension to universality. The system comprises a hierarchy of classes, a series of class definitions, a defined language for descriptive criteria, a new language for the class names, and a series of soil keys to allocate an unknown soil entity to its proper class. The classes and hierarchy are as follows:

Orders. Ten orders encompass all soil variation. They are distinguished by profile features which are the marks of the broad genetic processes dominant in shaping the characters of soils. The orders and their general characteristics are:

Entisols	recently formed soils
Vertisols	shrinking and swelling clay soils
Inceptisols	soils without contrasting horizons
Aridisols	soils of arid regions
Mollisols	soils with mull humus
Spodosols	soils with iron and humus *B* horizons
Alfisols	soils with clay *B* horizons and high base status
Ultisols	soils with clay *B* horizon and low base status
Oxisols	sequioxide-rich highly weathered soils
Histosols	peats

The orders and suborders may be seen to have descended (with a certain degree of rationalization) from the great soil groups of Baldwin *et al.* (1938). Entry to the classification system is by the key to orders, and this leads by other keys to suborders, great groups, subgroups, and finally to families and series. At each categorical level there is a definition and description of each of its classes.

Suborders. The orders are divided into 47 suborders which are subdivisions on characteristics of broad genetic significance, e.g. variants with poor drainage (five suborders), soil-moisture variants (8), soil-temperature variants (7), variants in material, bleached variants etc. The udalfs are an example of a suborder. They are soils of the order alfisol but with an 'udic' moisture regime.

Table 4.1 describes subdivisions of the order alfisols through the udalf suborder, to the 18 subgroups of the hapludalf group to provide a sample of the soil keys and classes of *Soil taxonomy.** It will be noted that the keys are very explicit and detailed. For many of the differentiae, which are stated in few terms, an exact and lengthy definition is to be found elsewhere in the text. For example, soil-profile materials which are considered to represent the results of fundamental pedogentic trends are instituted as 'diagnostic horizons', named and defined, and then used throughout the system. Table 4.2 lists a selection of these. Many of the complex differentiae will be recognized as the

* Table 4.1 is quoted from *Soil taxonomy* (Soil Survey Staff 1975) by kind permission of the Chief, Education and Publication Branch of the Soil Conservation Service of U.S.D.A.

TABLE 4.1.

Key to the classification of some alfisols

After the Histosols, Spodosols, Oxisols, Vertisols, Aridisols, Ultisols, and Mollisols have been separated, the Alfisols key in as

soils that:

1. Have an argillic or natric horizon but no fragipan,

or 2. Have a fragipan that

(a) is in or underlies an argillic horizon,

or (b) meets all requirements of an argillic horizon,

or (c) has clay skins > 1 mm thick in some part

Order . . . *Alfisols*

Alfisols are divided into suborders according to:

Alfisols that have an aquic moisture regime or are artificially drained and that have characteristics associated with wetness, namely, mottles or iron manganese concretions > 2 mm in diameter, or chroma of 2 or less immediately below any A_p horizon or below any dark A_1 horizon in which the moist colour value is less than 3.5 after the material is rubbed, and one of the following:

1. Dominant chroma of 2 or less in coatings on the surface of peds and mottles within peds of the argillic horizon, *or* a dominant chroma of 2 or less in the matrix of the argillic horizon and mottle of higher chroma;
2. If there are no mottles in the argillic horizon, a dominant chroma of 1 or less

Suborder . . . *Aqualfs*

Other Alfisols that have either:

1. A frigid temperature regime but do not have a xeric moisture regime:
2. A cryic temperature regime

Suborder . . . *Boralfs*

Other Alfisols that have one of the following:

1. An ustic moisture regime
2. An epipedon which is both massive and hard or very hard when dry and a moisture regime that is aridic but marginal to ustic.
3. Within a depth of 1.5 m of the surface or within a depth of 50 cm below the base of the argillic horizon, a calcic horizon or soft powdery lime in spheroidal forms or as coatings to peds or disseminated in particles of clay size and a moisture regime that is udic but marginal to ustic.

Suborder . . . *Ustalfs*

Other Alfisols that have one of the following:

1. A xeric moisture regime; or
2. An epipedon that is both massive and hard or very hard when dry and a moisture regime that is aridic but marginal to xeric

Suborder . . . *Xeralfs*

Other Alfisols that have a udic moisture regime

Suborder . . . *Udalfs*

In their turn Udalfs are subdivided into great groups:

Udalfs that have an agric horizon

Great group . . . *Agrudalfs*

Table 4.1 *continued*

Other Udalfs that have a natric horizon

Great group . . . *Natrudalfs*

Other Udalfs that

1. Do not have a continuous albic horizon above the argillic horizon,
2. Have a broken upper boundary of the argillic horizon,

and 3. Have discrete nodules in the argillic horizon that range from 2.5 to 5 cm to about 30 cm in diameter: exteriors of nodules are enriched and weakly cemented or indurated with iron and have a redder hue or stronger chroma than interior of nodules

Great group . . . *Ferrudalfs*

Other Udalfs that have tongues of albic material in the argillic horizon and do not have a fragipan

Great group . . . *Glossudalfs*

Other Udalfs that have tongues of albic material in the argillic horizon and have a fragipan

Great group . . . *Fraglossudalfs*

Other Udalfs that have a fragipan

Great group . . . *Fragiudalfs*

Other Udalfs that

1. Have mean summer and mean winter soil temperatures at a depth of 50 cm that differ by 5 °C or more;
2. Do not have a lithic or paralithic contact within 1.5 m of the soil surface;
3. Have clay distribution such that the percentage of clay does not decrease by as much as 20 per cent of the maximum within a depth of 1.5 m from the soil surface or the horizon in which the clay decreases either has > 5 per cent plinthite by volume or has skeletans or other evidence of clay eluviation; and
4. Have one or more of the following:
   (a) Hue redder than 10YR and a chroma more than 4 dominant in the matrix in at least the lower part of the argillic horizon
   (b) Hue of 2.5YR or redder and value, moist, of less than 4 and value, dry of less than 5 throughout the major part of the argillic horizon;
   (c) Many coarse mottles that have a hue redder than 7.5YR or chroma more than 5 or both

Great group . . . *Paleudalfs*

Other Udalfs that have an argillic horizon that has throughout its thickness a hue redder than 5YR, a colour value, moist, of less than 4, and a colour value, dry, no more than 1 unit higher than the value, moist.

Great group . . . *Rhodudalfs*

Other Udalfs that have mean summer and mean winter soil temperatures that differ by < 5 °C at a depth of 30 cm or a lithic or a paralithic contact whichever is shallower

Great group . . . *Tropudalfs*

Other Udalfs Great group . . . *Hapludalfs*

Hapludalfs are divided into subgroups:

continued. . . . . .

Table 4.1 *continued*

Typic Hapludalfs are Hapudalfs that

(a) do not have an abrupt textural change if there are mottles in the upper 25 cm of the argillic horizon;

(b) do not have a layer in the upper 75 cm that has texture finer than loamy fine sand, that is as much as 18 cm thick, that has a bulk density (at 1/3 bar water tension) of 0.95 g per cubic centimetre or less in the fine earth fraction, and that has either of the following:

(1) A ratio of measured clay to 15-bar water (percentages) of 1.25 or less; or

(2) A ratio of CEC (at pH near 8) to 15-bar water of > 1.5 and more exchange acidity than the sum of bases plus KCl-extractable aluminium;

(c) Have an argillic horizon that

(1) If the upper boundary is < 50 cm below the soil surface, does not have mottles that have chroma of 2 or less in the upper 25 cm if it is saturated with water within that depth at some time when the soil temperature is 5 °C or higher; or

(2) If the upper boundary of the argillic horizon is deeper than 50 cm, does not have mottles that have chroma of 2 or less within a depth of 75 cm below the soil surface;

(d) Have, within 50 cm of the surface, texture finer than loamy fine sand in some subhorizon;

(e) Do not have interfingering of albic materials and albic materials surrounding some peds in the upper part of the argillic horizon;

(f) Do not have a lithic contact within 50 cm of the soil surface;

(g) Have an A_p horizon that has a colour value, moist, of 4 or more or has a colour value, dry, of 6 or more (crushed and smoothed) or the upper soil to a depth of 18 cm, after mixing has these colours;

(h) Have an argillic horizon that is continuous horizontally and continuous vertically for at least the upper 20 cm of its thickness and that has a texture finer than loamy fine sand

(i) Have base saturation (by sum of cations) of 60 per cent or more at a depth 1.25 m below the top of the argillic horizon, or 1.8 m below the soil surface, or immediately above a lithic or paralithic contact, whichever is least;

(j) Do not have the following combination of characteristics:

(1) Cracks at some period in most years that are 1 cm or more wide at a depth of 50 cm and at least 30 cm long in some part and that extend to the surface or to the base of the A_p horizon;

(2) A coefficient of linear extensibility (COLE) of 0.09 or more in a horizon or horizons at least 50 cm thick and a potential linear extensibility of 6 cm or more in the upper 1 m of the soil or in the whole soil if a lithic or paralithic contact is deeper than 50 cm but shallower than 1 m; and

(3) More than 35 per cent clay in horizons that total > 50 cm in thickness within the control section; and

(k) Do not have albic materials that constitute as much as 5 per cent of the volume of any subhorizon of the argillic horizon.

Subgroups:

Typic Hapludalfs—as defined above.

Albaquic Hapludalfs are like Typic Hapludalfs except for *a* and *c* with or without *g*

continued......

Table 4.1 *continued*

Albaquultic Hapludalfs are like Typic Hapludalfs except for *a*, *c*, and *i*
Andeptic Glossoboric Hapludalfs are like Typic Hapludalfs except for *b* and *e*, and their mean annual soil temperature is lower than 10 °C
Aquic Hapludalfs are like Typic Hapludalfs except for *c*
Aquic Arenic Hapludalfs are like Typic Hapludalfs except for *c* and *d* with or without *g* or *i* or both
Aquic Lithic Hapludalfs are like Typic Hapludalfs except for *c* and *f*
Aquollic Hapludalfs are like Typic Hapludalfs except for *c* and *g*
Aquultic Hapludalfs are like Typic Hapludalfs except for *c* and *i* with or without *g*
Arenic Hapludalfs are like Typic Hapludalfs except for *d* with or without *g* or *i* or both
Glossaquic Hapludalfs are like Typic Hapludalfs except for *e* and *c* or for *e*, *c*, and *g*
Glossic Hapludalfs are like Typic Hapludalfs except for *k* with or without *e* or *g* or both, and the mean annual soil temperature is 10 °C or higher
Glossoboric Hapludalfs are like Typic Hapludalfs except for *e* or for *e* and *g* and their mean annual soil temperature is lower than 10 °C
Lithic Hapludalfs are like Typic Hapludalfs except for *f* with or without *g*
Mollic Hapludalfs are like Typic Hapludalfs except for *g*
Psammentic Hapludalfs are like Typic Hapludalfs except for *h* with or without all or any of *d*, *g*, or *i*
Ultic Hapludalfs are like Typic Hapludalfs except for *i* with or without *g*
Vertic Hapludalfs are like Typic Hapludalfs except for *j* with or without any or all of *a*, *c*, *g*, or *i*.

morphological indicators of certain soil processes and as indicators of genetic factors. The system could be recognized as genetically motivated even though the differentiae are expressed in morphological terms. This implies absolute reliance on a perfect transposition from genetic to morphologic terms.

Great groups. The suborders are divided into 185 great groups on the basis of characteristics affecting the whole profile such as contrasting material (38), minimal-profile expression (21), frozen ground (18), ancient soils (15), tropical conditions (14), other temperature and moisture variants (12), natric horizons (12), hardpan (11), fragipan (10), plinthite (9) etc. It is not easy to see any contrast in scope or scale between these differentiae and those between the suborders above.

Subgroups. The great groups are divided into 970 subgroups. The central concept (typic) of each group is recognized, and intergrade or extragrade subgroups are recognized and defined on one or more deviant characteristics which do not affect the whole soil. Subgroups

TABLE 4.2.
List of diagnostic horizons and formalized class differentiae used in 'Soil Taxonomy'

Mollic epipedon	– soft, mol	*Moisture regimes*
Anthropic epipedon	– man-made	aquic
Umbric epipedon	– dark	udic
Histic epipedon	– peaty	ustic
Plaggen epipedon	– sod	xeric
Ochric epipedon	– pale	aridic
Argillic horizon	– clay illuvial	or torric
Agric horizon	– agricultural layer	
Natric horizon	– sodium layer	
Sombric horizon	– illuvial humus	
Spodic horizon	– podzol layer	
Placic horizon	– thin iron pan	
Cambic horizon	– exchange layer	
Oxic horizon	– thick iron	
Duripan horizon	– hard pan	*Temperature regimes*
Fragipan horizon	– brittle pan	pergelic
Albic horizon	– bleached layer	cryic
Calcic horizon	– lime layer	frigid
Gypsic horizon	– gypsum layer	mesic
Petrocalcic horizon	– lime-rock layer	thermic
Salic horizon	– salt layer	hyperthermic
etc.		

are named by adjectives before the group name, e.g. albic glossic natraqualf.

Families. So far there are some 4500 recognized soil families in the United States. This is a pragmatic category and has no apparent pedogenetic reference. Any of the following are criteria for classes: particle-size classes (7 or 11), mineralogy classes (17), calcareous and reaction classes (4), soil-temperature classes (2 × 4), soil-depth classes (2), soil-slope and shape classes, soil-consistence classes (2), classes of coating (2), and classes of cracks.

The family is named by the addition of adjectives to the subgroup name; for example, *coarse silty over clayey, mixed mineral, mesic, albic glossic natraqualf.*

Series. This is the lowest category and like the family is a pragmatic grouping of soils on any property variation which might influence use and behaviour. These are the soil surveyor's primary groupings of soil entities, and some 10 500 have been recognized in the United States. Though pragmatic they must comply with all the differentiae in all the categoric levels above them. A series class must not straddle or cross any soil-class boundary above it in the hierarchy, nor can it have more nor less range of variation than the polypedon by which the soil-series area is represented in the field.

The soil series name is a place name from its area of origin, usually awarded and defined during a soil survey. The series area, called the polypedon, is the basic soil-mapping entity and may be subdivided into phases on variations in slope, texture, stoniness etc. that affect its use where these have not been isolated by the system already. The polypedon is the connecting link between the classes of the system and the areas of land in which soils are manifest and where they are studied and used.

Comment. The greater part of the system consists of galleries of rationalized classes. These have been created by a downward elaboration of the original conceptual classes by innumerable taxonomic chops. Some of the principles governing the success of taxonomic chops have been discussed already (p. 19). Unless the chops are well informed they can easily produce classes of negligible predictive value, or even lead to a loss of information in the process of classification. There is no clear evidence that these principles have been followed in the production of *Soil taxonomy*, so its value as a basis for predictions needs confirmation.

The differentiae of the system are mostly diagnostic horizons and similarly defined materials, defined in their turn on soil morphologic attributes, laboratory analytic data, and on climatic regimes of moisture and temperature. The morphologic attributes of colour, texture, consistence, structure, and accessory materials are easily accessible, but the use of some of the other differentiae that depend on laboratory analyses is likely to depend on extrapolations from a few well studied sites to the whole of a survey area.

The main impression of the system is its detailed interlocking structure, its wealth of definition, and the unambiguity of its allocation of new soil entities to soil classes. But its very comprehensiveness makes it hard to find out *why* it is structured as it is, how it may be tested, and on what presumptions it is based. In the Foreword to *Soil taxonomy*, Johnson says that the important question at each step in the development of the system was 'do these groupings permit us to make precise predictions of soil behaviour?' We are left to assume that such predictions follow from a known association between soil-use factors and the differentiae which are used to create the classes. Such fine and numerous differentiae would only be justified if it were known that there was a high degree of orderliness in the data and if its structure were known. If the soils of the United States and all other areas to which *Soil taxonomy* is being applied do not possess a high state of orderliness, then the great detail in keying and in the definition of differentiae is wasted, and it creates no more than 'key-characterized classes'. In spite of the present prolixity of the system, its development apparently established no principle to limit its further growth by adjustment to fit new soil-data pools. In a universe of limited orderliness *every* combination of characteristics is possible and, if adjustments to the system are not restrained by some principle, its prolixity will eventually be overwhelming. This is a dangerous position for a classification system to be in.

Another disadvantage of a fixed hierarchy is the restraint it imposes on the series units that arise during soil survey. Ideally a series should be developed by the soil surveyor so that it fits the modes of local occurrence of profile variation and boundaries. If the series is part of a fixed hierarchy, then its range of variability cannot cross the differential point of any class above it in the system, and all the differentiae of all the classes and categories above must apply to it also in addition to its own intraseries differentiae. This is likely to impose compromises and inaccuracies at the series level where accuracy should be at its sharpest. The institution of a hiatus between the series and the rest of the system would overcome this want of flexibility. The very principle of fixity in the hierarchical structure may be queried; it tends to give specificity to the high-level classes where specificity is a chimera, and it does this at the expense of specificity in the low-level classes where it is most desired and most possible.

Soil taxonomy is unusual in the high proportion of its class differentiae which are climatic attributes. Both temperature (six classes) and moisture (five classes) regimes are treated as profile attributes. However, they are usually assessed on meteorological data when a new soil is being allocated to the system. Because data such as the moisture regime of a soil profile are only rarely determined in this manner, meteorological figures are generally used as surrogates for the profile attributes and in some places meteorological boundaries become soil boundaries on the ground.

Russian soil classification (Gerasimov and Glazovskaya 1960)

Although modern soil science began in Russia, soil classifications there have been more consistent than other national systems and showed none of the perturbations that affected other classifications in the 1950s. Russian soil classification and soil science generally emphasizes soil dynamics (Basinsky 1959); i.e. the dynamic role of vegetation, organic decomposition, leaching, mineral weathering, and the moderating effect on these of environmental conditions such as climate, landscape drainage, and relief. With this emphasis has gone a relative diminution in the significance of particular features of the soil profile. The soil classification of Gerasimov and Glazovskaya then is geographically oriented (indeed 'zone' and 'soil class' are two facets of the one dynamic system in the Russian approach) and has the following categories:

(a) genetic type
(b) subtype
(c) species
(d) subspecies, variety, genus, and subvariety.

Gerasimov and Glazovskaya remark [p. 179] on the *type* (the highest categoric level) that 'the principle criterion of unity of origin, as well as identity of process of transformation and translocation of substances, in a group of soils comprised in a common type is similarity of make-up of the general profile, expressed by the presence of the same system of interrelated horizons. A complementary criterion of the soils' unity within the type is the similarity of conditions of soil formation'. They say further [p. 180] that 'the notion of *subtype* introduces greater detail into the concept of type by singling out groups of soils where the general features of a soil type are supplemented by peculiar traits of the genetic profile due to particular conditions of soil formation'.

The concept of *species* is used to 'designate quantitative stages of development in soil-forming processes characteristic of the soil type'.

It is not clear that the system is hierarchical below this level. The *genus* distinguishes within subtypes variations dependent on parent rock including saline manifestations. *Varieties* and *subvarieties* are especially called for in detailed soil survey and mapping, the former referring particularly to classes of soil texture and the latter to variations in a genetic character that are due to different petrographic composition–'Thus one group of taxonomic subdivisions used in

Soviet pedology (genetic soil type, subtype, and species) defines mainly those genetic soil properties which develop autonomously, whereas another group of such subdivisions (genus, variety, and subvariety) defines residual or lithological peculiarities inherited by a soil from its parent material'. [*loc. cit.*, p. 18].

The *genetic types* occur in zones of the same name. These are as follows: Tunda-Glei soils; Podzolic soils of Taiga Forests; Podzolic-marshy soils; Marshy soils; Turfy soils; Brown and grey forest soils; Cinnamon and grey-cinnamon soils; Chernozem and meadow-Chernozen soils of the steppes; Chestnut and meadow-chestnut soils of the dry steppes; Brown desert-steppe and grey-brown desert soils; Solonchak, Solonetz, and Solod; Takyr and Takyr-like soils; Serozem and meadow-Serozem soils; Krasnozems and Zheltozems; Soils of mountain regions. Each soil type is given a well rounded description covering its geographic environment, climatic regime, natural vegetation, seasonal organic matter regime, the dominant soil processes (e.g. leaching, humification, mineralization, mineral weathering), the main profile features, and the significant role of specific horizons. The class is described largely in terms of its typical or central concept and little or no attention is given to defining class boundaries.

As an example of the genetic breakdown of a type, Table 4.3 presents subdivisions of the soil type 'Podzolic soils of the Taiga Forest'.

TABLE 4.3.
Subdivision of the soil type: Podzolic soils of the Taiga Forest

Subtypes –	Typical	
	Glei-podzolic soils	
	Podzolic-illuvial humus	
	Turfy-podzolic	
Species –	Weakly podzolic	determined by thickness of A_2 horizon; apply to all subtypes
	Medium podzolic	
	Strongly podzolic	
Subspecies –	classes grouped on thickness of humus accumulation	

After Gerasimov and Glazovskaya 1965.

Comment. The Russian classification places most emphasis on soil process and on the pedogenetic functions of the environment. Yet the routine identifying of soil entities, as in soil survey, must rely on profile morphology because determinations of process and genesis involve numerous analyses and calculations which must always be too few. The soil surveyor would be driven to extrapolate widely any discovered association between process and morphology and to extrapolate from researched sites by means of profile morphology, colour, structure consistence, and arrangement of horizons.

Within its own particular balance of interests the Russian classification has a simple structure. It consists of a gallery of conceptual classes (the types), and the rationalized elaboration of this has created the three lower categories of subtypes, species, and subspecies. However, these have not yet been tied up into a firm hierarchy by a comprehensive system of definitions. Although the rationalized elaboration has gone far enough to assist the marshalling of data and ideas about soil variation, in a general way it has not been carried far enough to establish keys and to define classes in a comprehensive way. There is not much definition worked into the system, so we would expect that it suffers ambiguity in identification and miscellaneousness of allocation, which Leeper and others complained of in the 1950s.

Northcote's system for Australia (1960)

This system merits attention because of its unusual form; it was initiated by a key (Northcote 1960, 1971) and has no gallery of conceptual classes as its foundation. Northcote uses a simple bifurcating key based on simple defined soil-profile characteristics, most of which can readily be determined in the field. The first division is on the nature of the soil-texture profile—uniform texture, gradual texture increase, or abrupt texture increase—and the two differentiating points are defined. The second division in the key is different for each division as shown in Table 4.4. There are six levels of division in the key, and Table 4.5 presents an example of the lower four. Table 4.6 summarizes the kind of soil characteristic used as differentiae at each of the categorical levels. Note that texture, colour, and consistence are each used as differentiae several times over. Once identified by the key, a soil entity is allotted a coded identity which describes it in terms of the differentiae used. All of the positions in the key should be regarded as pigeon holes; they are defined on their borders, but frequency

TABLE 4.4.

Northcote's key – differentiae for divisions and subdivisions (adapted from Northcote 1971)

All soils – *divisions*		Code
	1. Uniform texture profile	U
	2. Gradational texture profile	G
	3. Abrupt texture profile	D (duplex)
Breakdown of divisions		
U –	1. Coarse textured	Uc
	2. Medium textured	Um
	3. Fine textured not cracking	Uf
	4. Fine textured, cracking	Ug
G –	1. Calcareous throughout	Gc
	2. Not calcareous throughout	Gn
D –	1. Red clay *B* horizon	Dr – (See Table 4.5)
	2. Brown clay *B* horizon	Db
	3. Yellow grey clay *B* horizon	Dy
	4. Dark clay *B* horizon	Dd
	5. Gley clay *B* horizon	Dg

modes of different soils within their range are not indicated, nor can it be assumed that the median form is the most common.

Northcote *et al.* (1975) carried the system further by making a summary of the soils that had been found to occur in the system's pigeon holes during some ten years of soil-mapping experience all over Australia. It was found that 307 of the pigeon holes were occupied by enough soils to warrant the status of a class. This means that 307 out of 855 pigeon holes were relevant and the balance (548) were not relevant.

Whatever 'relevant' may mean in this context, it is possible to examine the soils that have found themselves in each of the 307 classes to see if they exhibit any consistency in characteristics over and above those used in keying them into the groups. Northcote *et al.* (loc. cit.) have done this, and they describe associated characteristics like landscape and land use as well as the modalities of both keying and non-keying soil properties. Unfortunately they have not put the modes and associations discovered to any test beyond personal observation.

Comment. Northcote has given no philosophical or conceptual reasons

TABLE 4.5.

Northcote's key – example of breakdown of subdivisions, sections, and classes (adapted from Northcote 1971)

Breakdown of subdivision Dr		*Sections*
A horizon with crust		Dr1
A horizon hard setting, *B* horizon whole coloured		Dr2
A horizon hard setting, *B* horizon mottled		Dr3
A horizon not hard setting, *B* horizon whole coloured		Dr4
A horizon not hard setting, *B* horizon mottled		Dr5
Breakdown of section Dr1		*Classes*
AA.	Peds evident in clayey *B* horizon	
	No *A*2 horizon	Dr1.1
	*A*2 horizon present, not bleached	Dr1.2
	*A*2 horizon sporadically bleached	Dr1.3
	*A*2 horizon conspicuously bleached	Dr1.4
AA.	Few if any peds in clayey *B* horizon (similar breakdown to Dr1.1 to .4)	
Breakdown of Class Dr1.1		*Principal Profile Forms*
BB.	No hardpan in or below the clayey *B* horizon	
	Acid soil reaction trend	Dr1.11
	Neutral soil reaction trend	Dr1.12
	Alkaline soil reaction trend	Dr1.13
BB.	Hardpan present in or below the clayey *B* horizon (similar breakdown to Dr1.11 to Dr1.13)	

for choosing to start a classification with a key, nor for choosing the differentiae that he has. We may assume, however, a desire to make an unambiguous sorting of soils and to use criteria which are simple and objective and do not involve preconceptions about the genetic importance of soil features nor depend on an interpretation of pedogenesis. Leeper (1954) suggested such objectives for a new classification. The avoidance of unproven assumptions between soil features and factors seems to have led to a choice of differentiae which have no known relevance at all, except that they allow the system to work freely. Furthermore, it is not yet known which associated properties

TABLE 4.6.
Characteristics used as differentiae in Northcote's key

Category	*Number of pigeon holes*	*Differentiae*
Divisions	3	Texture gradient to clay *B* horizon (three options)
Subdivisions	11	Texture (four options) Calcareousness (two options) Colour of clay *B* horizon (five options)
Sections	54	Pedologic organization, bleaching of A_2 (six options) Fabric of *B* horizon (four options) Consistence of *A* horizon, mottling in *B* horizon
Classes	271	Consistence, fabric, hard pan, colour, fabric of clay *B*, calcareousness, structure of clay *B*, presence of A_2
Principal Profile Forms	855	Calcareousness, colour, Soil-reaction profile Colour of clay *B* hardpan Laterite layer
	The listing is not exhaustive	

are grouped by the keying properties. Since the prediction of such associated properties is usually the *raison d'être* for the selection of keying properties, their identification vindicates the classification. The soil map of Australia, in which the mapping units are described according to the Northcote key (Northcote *et al.* 1975), does not show any obvious patterning of soils in relation to climatic trends or the main geological formations of the continent. This is mainly because acknowledged 'pedogenetic factors' are excluded from the differentiae of the key. The main patterning to the soil map is of soil texture and this is in keeping with the nature of the differentiae used. The main divisions of the key (U, G, and D) are not likely to give a structure to a map since any of them may occur adjacently in the same catena or hillside section.

The fact that the pigeon holes can be filled from soil-mapping experience does not prove that the system 'fits' the natural occurrence of soils. The situation is illustrated in Fig. 4.1 in which property *X* is

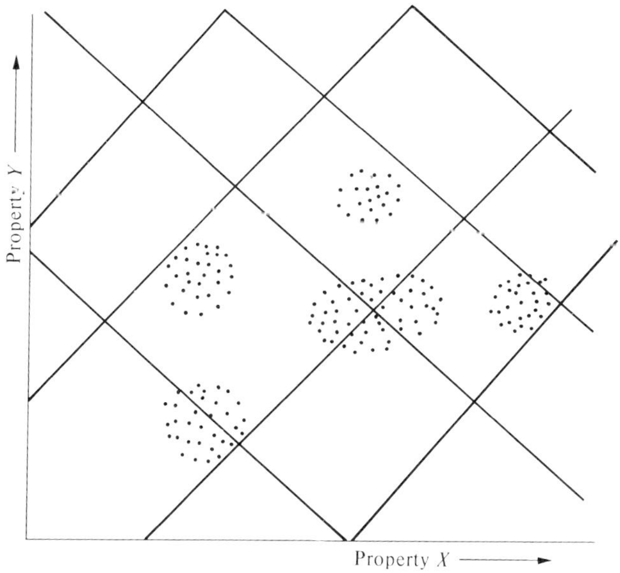

Fig. 4.1. Imaginary superimposition of the graticule of a key system over a possible set of natural soil entities plotted out in terms of their properties *X* and *Y*. The pigeon holes made by the key would collect varying numbers of soil entities, but these only roughly reflect the distribution of soils in property space. Some of the pigeon holes would gather bimodal and trimodal populations.

plotted against property *Y* for the soils of a region of interest. We assume that there is some natural clustering of data. If the graticule of the key were superimposed on the plot (as in the diagram), the fact that some pigeon holes had many entrants and some had few or none would not prove a correspondence between the natural distribution and the key. Any keying graticule could be expected to show some grouping of the data. Of more significance would be the fact that the contents of some of the pigeon holes may show bimodal distributions or only half a normal distribution.

The absence of a gallery of conceptual classes from a classification system deprives it of a full well rounded series of images. It is true that a set of images has now developed from some years of using the scheme, but these have been shaped to an unknown degree by the key itself.

The soil classification: 'A handbook of Australian soils' (Stace *et al.* 1968)

This system is unusual nowadays because it consists of little more than a gallery of conceptual classes and thus corresponds to only the initial stage in the development of most contemporary systems.

The system in the Australian handbook offers a single category of 43 great soil groups. These are grouped into seven classes at a higher but unnamed level by a simple rationalization, and the authors suggest that a lower category of families might be developed to bridge the interval between the great soil groups and the series of detailed soil surveys. Table 4.7 lists the great soil groups and the higher unnamed category and is copied from Stace *et al.* (1968) by permission of Rellim Technical Publications. The published system provides no definitions or descriptions of the classes of the higher category. No doubt they are self explanatory and in any case they appear to serve no purpose.

Each of the great soil groups is described under a great many headings. The presentation of the red-brown earth great soil group is typical (Stace *et al.* 1968, pp. 211–27) and includes a colour picture of a typical profile; a full description of a typical profile (including a statement of the main variant forms and the range of variation); a description of the micromorphology of a typical profile; a statement of the soils' occurrence, geographic distribution, and landscape associations; a statement of the genesis of the red-brown earth profile; a statement on associated land use and cropping; and reference to the specific advantages and problems of this soil. There follow specific descriptions of five selected representative profiles, together with their chemical, mineralogic, and physical analyses, and micromorphologic descriptions with pictures and genetic interpretation. There is no statistical analysis of the five sets of data.

This treatment is typical of all the classes in the system in presenting a centroid description covering a great many properties and attributes of the soil in its environment, but because it does not define any class boundaries or identify any essential attributes, there is no soil key. Classes of this sort represent the condensation of a vast amount of experience, and it would be of great value to identify a new soil to its place in such a system. But the allocation of new soils to their place in the system is sometimes uncertain or miscellaneous.

TABLE 4.7.

Great soil groups of the Australian Handbook (Stace et al. 1968)

Higher category	*Great soil groups*
1. No profile differentiation	1. Solonchaks
	2. Alluvial soils
	3. Lithosols
	4. Calcareous sands
	5. Siliceous sands
	6. Earthy sands
2. Minimal profile development	7. Grey, brown, and red calcareous soils
	8. Desert loams
	9. Red and brown hardpan soils
	10. Grey, brown, and red clays
3. Dark soils	11. Black earths
	12. Redzinas
	13. Chernozems
	14. Prairie soils
	15. Wiesenboden
4. Mildly leached soils	16. Solonetz
	17. Solodized solonetz and solodic soils
	18. Soloths
	19. Solonized brown soils
	20. Red-brown earths
	21. Non-calcic brown soils
	22. Chocolate soils
	23. Brown earths
5. Soils with predominantly sesquioxide clay minerals	24. Calcareous red earths
	25. Red earths
	26. Yellow earths
	27. Terra rossa soils
	28. Euchrozems
	29. Xanthozems
	30. Krasnozems
6. Mildly to strongly acid and highly differentiated	31. Grey-brown podzolic soils
	32. Red podzolic soils
	33. Yellow podzolic soils
	34. Brown podzolic soils
	35. Lateritic podzolic soils
	36. Gleyed podzolic soils
	37. Podzols
	38. Humus podzols
	39. Peaty podzols
7. Dominated by organic matter	40. Alpine humus soils
	41. Humic gleys
	42. Neutral to alkaline peats
	43. Acid peats

Comment. Although Leeper criticized this style of classification in its earlier form (Stephens 1953) and although it has the weaknesses he identified, it has persisted as an interesting survival and demonstrates that there is a role for a gallery of conceptual classes at certain stages in the use of soil knowledge and in the learning of soil science. To convert it into a more elaborate defined system would involve deciding which are the more important attributes of each of the classes and which *values* of each attribute should be adopted as cut-offs between each class and the next ones. It would then be necessary to decide what subdivisions into classes of lower category were justified, the cut-offs between these, and so on. All this would be difficult.

FitzPatrick's system

FitzPatrick's (1971) classification is most unusual; it seems to be based on the philosophy that soils are a continuous mantle in the landscape varying in place, depth, and time so that 'the situation defies classification' (loc. cit., p. 106) and that we should avoid bringing pedogenetic presumptions into a classification. His classification then is based on minimum assumptions.

The stages of the system from the individual soil upward are:

Horizon. The logical point of departure in this system is the recognition of classes of horizon material. FitzPatrick recognizes 77 classes of horizon. These are described and defined and keys are provided for the identification of unknown horizons. Table 4.8, which is quoted by kind permission of FitzPatrick and Oliver and Boyd, illustrates his treatment of kastanon, krasnon, kuron, and limon horizons. Horizon classes also receive a well rounded description in terms of morphology, analytical properties, thin-section morphology, genesis, variability, associated horizons, and distribution.

Pedounit. A pedounit is an actual combination of horizons encountered in the field. It is described by listing the coded symbols of the horizons present and the thickness of each in order from the soil surface downward. Pedounits have no *a priori* status, and are not presumed to be genetically inter-related.

Group. A group is a generalized collection of pedounits grouped together because of their similarity and recognized for the time being as a useful class for soil mapping or other purposes. As a mapping unit a group is equivalent to the series and it is given a locality name. It is described and defined for the soil-survey project, but it has no general status beyond that. It seems to be implicit from this stance that there could also be miscellaneous entries on the map corresponding to pedounits which do not fall into any of the apparent groups.

Subclass. A subclass is a category above the series (group) level and consists of series grouped together because of their similarity in two or three major horizons. The subclasses, like the series, have an *ad hoc* status and are not defined or keyed. This level is not yet developed to any great extent.

Class. The class is the most generalized category in the system. It consists of subclasses grouped together on the basis of a prominent horizon or a unique combination of certain horizons. The classes also have only *ad hoc* status. There is no definition or key to classes but each class has a full and broad description covering general characteristics, macro- and micro-morphology, analytical data, genesis, principal variations in the properties of the class, parent material, climate, vegetation, topography, age, common pedounits, distribution, utilization, derivation of names, and approximate equivalent names.

FitzPatrick describes 33 soil classes. Their general categorical level is about that of the suborder of the US soil taxonomy and their names are listed in Table 4.9.

This system proposes a binomial system of naming in which a class name is followed by a series or group name, e.g. Podzol Corby.

Comment. The peculiarities of FitzPatrick's system are grounded in the problem of the non-entity of soils. Classification is a grouping of entities. How are we to classify soils when they have no natural entities?

He meets the problem by allowing populations of soil data to form their own groups on their perceived similarity, and he refuses to make any one experience of this kind a general rule for the future.

In spite of its unusual character, FitzPatrick's system shares with other systems

(a) a level of generalization at the soil surveyor's level where soil entities are grouped into primary classes;

(b) another higher level of soil generalization where 'all soils' on

TABLE 4.8.

Example of FitzPatrick's (1971) descriptions and definitions of four horizons. Differentiae are underlined.

Horizon	Kastanon	Krasnon	Kuron	Limon
Name and symbol	Kt	Ks	Ku	Lm
Position	Upper	Upper or middle	Upper	Middle or lower
Colour	Dark brown Very dark grey	Red, brownish red*	Black, very dark brown	Grey to white
Thickness (cm)	35–50	25–> 3m	10–25	3–100
Texture	Loam, silt	> 30% clay†	Loam, sand	Silt
Stones and boulders	Usually absent	Absent to rare	Absent to rare	Absent
Structure	Crumb or granular to prismatic with depth	Massive, blocky	Granular, crumb, subangular blocky	Massive
Consistence	Firm, friable	Friable, firm, hard, plastic	Friable, greasy	Firm, plastic
OM (%)	2–5*	5	5–20	5
C/N	8–12	8–12	12–20	15–40
pH	7.0–8.5	4.5–5.5	4.5–6.6	7.5–9.0
CEC (me%)	10–40	< 15 (clay)	20–50	–
BS (%)	100	< 15	10–30	–
Carbonates (%)	Pseudomycelia and concretion in lower part. 0–60 from parent material†	Absent	Absent	> 30
Salts (%)	< 0.5	Absent	Absent	Absent

Clay minerals	Mica F-A Montmorillonite F-A	Kaolinite F-D Gibbsite F-D Goethite F-D Hematite < F Mica < F	Allophane A-D Halloysite < F	
Weatherable minerals (%)	20–60	< 5	5–80	Variable
Thin section morphology	Numerous faecal pellets and worm casts. Isotropic. In lower part, surfaces covered with calcite crystals also needles and concretions	Abundant random very small domains	Isotropic matrix	
Associated horizons Above:	Litter	Tannon, mullon, flambon	Litter	Amorphon, pseudofibron
Below:	Calcon, gypson, parent material	Pesson, veson, chemically weathered rock	Andon, parent material	Parent material
Intergrades and variants	(KtCh) (KtBu)	(KsRf) (KsRo)Ksh	(KuMu)	
Illustrations:				
Plates		I11a	Ib	
Figs.		141, 143		
Other distinguishing features	* 15–20 kg OM per m² † When derived from parent material Ktc.	* 10R and 2.5YR † Maximum content of clay in middle of horizon – not due to translocation. When higher Krh. Silt: clay ratio < 0.15 on crystalline rocks < 0.2 on sediments		Permanently saturated with water

Table 4.8 *continued*

Horizon	Kastanon	Krasnon	Kuron	Limon
Genesis	Rapid breakdown and incorporation of organic material in a dry environment	Hydrolysis and clay mineral formation in a humid tropical climate	Decomposition and blending or organic mineral material	Accumulation of precipitated calcium carbonate or skeletons of aquatic animals on the bottom of a lake
Derivation	R. *Kastano* = chestnut	R. *Krasni* = red	J. *Kuroboku* = dark	I. *Limus* = slime

TABLE 4.9.
FitzPatrick's soil classes

Primosols	Chernozems	Luvosols	Serozems
Rankers	Clamosols	Peat	Solods
Altosols	Cryosols	Pelosols	Solonchaks
Andosols	Flavosols	Placosols	Solonetzes
Arenosols	Gelosols	Planosols	Subgleysols
Argillosols	Halosols	Podzols	Supragleysols
Brunizems	Kastanozems	Rendzinas	Thiosols
Burozems	Krasnozems	Rossosols	Vertisols
			Zheltozems

a national or world scale are comprehended and put into classes. All soil-classification systems contain these two working levels; they differ in the status given to the two levels and how the gap between them is to be bridged. FitzPatrick believes that there should be no firm bridging structure of hierarchic or rationalized elaboration between them because this would presume that we know more about soil variation than we do. He allows the *found* variation, the frequency modes and discontinuities of occurrence in the data before us, to determine the groupings that we make, and any grouping is made without committing ourselves to making it a general rule.

FitzPatrick's stage of definition and keying is at the level of the horizon (i.e. *before* the soil profile has been dealt with), and consequently there are no presumptions because there are no definitions of what the profile, series or class shall be, before the weight of evidence reveals what each of these is to be in each particular study.

This seems to conform to the logic of the situation. The series should not be defined last (after order, suborder, great group, subgroup, and family); it should be defined first because it is to be the mapped class and its class boundaries need to fit the boundaries discernible in the landscape. It is at this level that precision is required and not at the higher categories of classification.

Kubiena's system for the soils of Europe (1953)

This is a popular classification and merits attention as a well-developed example of the older style of classification system (Kubiena 1953). It has six categories: Divisions (three); Classes (ten); Types (20); subtypes (52); Varieties; and Subvarieties. Two of the three Divisions are

TABLE 4.10.
Soil classes of the division of terrestrial or land soils

1. Terrestrial raw soils	6. Bolus-like silicate soils
2. Ranker-like soils	7. Latosols
3. Rendsina-like soils	8. Brown earths
4. Steppe soils	9. Pseudogley soils
5. Terrae Calxis	10. Podsols

TABLE 4.11.
Soil types of the division of terrestrial soils

1.1	Rawmark	4.5	Para-chemosem
1.2	Yerma	4.6	Para-serrsem
1.3	Syrosem	5.1	Terra
2.1	Ranker	6.1	Braunlehm
3.1	Eurendsina	6.2	Rotlehem
3.2	Para-rendsina	7.1	Latentic roterde
4.1	Serosem	8.1	Braunerde
4.2	Burosem	9.1	Pseudogley
4.3	Kastanosem	10.1	Semi-podsol
4.4	Chernosem	10.2	Podsol

underwater soils, and flooding and groundwater soils. We will examine the third Division—terrestrial or land soils.

Terrestrial or land soils are divided into the ten *classes* of Table 4.10. The classes are divided into 20 *types* which are listed in Table 4.11. The types are equivalent to the great soil group level of generalization and are the gallery of conceptual classes on which the system is based. There are 52 *subtypes* and these are divided again into *varieties* and *subvarieties*. The subvarieties are still above the level of the series in detailed soil survey. The higher categories of classes and divisions are logical condensations of the types, and the lower categories of subtypes and varieties are rationalized subdivisions of the types.

The system includes a full description of each type in terms of soil profile, humus form, chemical characteristics, dynamics (soil processes), micromorphology, variability, and the main deviant themes of variation, geographic distribution, and associations. Subtypes and varieties are also given a full description, but classes are not (as befits their status as rationalized entities).

A striking feature of Kubiena's system is its five keys, of which the fifth is presented as the most exact. The other keys provide alternative means of allocating an unknown soil to its type and variety in the system. Kubiena remarks that, although the character of a soil is the total of all its characteristics and attributes, it can be differentiated from the other soils with which it is associated by the one characteristic which is critical in their association; thus full knowledge of the soil scene enables an operator to make effective and economical keys. Tables 4.12 and 4.13 present extracts from key number 5 to show the kinds of differentiae used. Note that the differentiae are mainly broadly stated morphologic characteristics together with some environmental differentiae.*

Kubiena's book is formed around his soil keys and it is intended that the keys should provide the point of entry of an unknown soil and should lead to the description of its soil class. The match of the soil to the description of the class is the test of the operation of the key.

Comment. The success of Kubiena's system is probably due to the good match between his types—the gallery of conceptual classes in this case—and the soil experience of those seeking to use it. After 24 years his type concepts are still useful. Another reason for the early success of the system was probably its widespread use of soil keys in a time when soil keys were rarely used. Nevertheless, by modern standards the keys are imprecise; they operate on the central concepts and do not define interclass boundaries.

The keys also use criteria of a diverse kind—morphological characteristics, broad environmental, and soil-process attributes. The use of these could lead to ambiguity. In many parts of the key the differential is one characteristic only, a situation needing reconciliation with the polythetic descriptions of the classes thus identified. The presumption of a high degree of orderliness in the soil data is pervasive but must be queried.

* These tables are copied from Kubiena (1953) by kind permission of the Director Consejo Superior de Investigaciones Scientificas Madrid.

TABLE 4.12.

Example of Kubiena's (1953) key number 5 (C): key for the determination of classes (p. 143)

1.	Soils without macroscopically distinguishable humus horizons. CA. Terrestrial raw soils.	
1*.	Soils with macroscopically distinguishable humus horizons.	2
2.	Soils with distinct humus horizon, however, without *B* horizons, i.e. the humus horizon lies directly on the parent material or on a layer broken down only by mechanical weathering, not changed essentially by chemical weathering (AC soils).	3
2*.	Soils with *B* horizons, i.e. with subsoil of brown, ochre yellow to red colour.	5
3.	Soils with generally uniform development on broad areas in the steppe regions or summer dry cultivated areas, occurring only on loose parent material. CD. Steppe soils.	
3*.	Soils occurring generally on hard parent rocks or outside summer dry areas on loose parent material.	4
4.	Soils calcareous or containing chalk humates, occurring on chalk, dolomite, gypsum, or chalk-silicate rocks. CC. Rendsina-like soils.	
4*.	Soils low in chalk and containing no chalk humates, occurring almost only on silicate and siliceous rocks, low in chalk, very rarely on chalk-silicate rocks. CB. Ranker-like soils.	
5.	Soils with the appearance of gley soils, i.e. with light grey, dense structured horizons (frequently only in the top soil under the humus layer), but already distinguished from them by a strong alternation of wetting and drying in which the drying phase usually predominates. In general they show small concretions of ferric hydroxide in the top soil and strongly flecked to marbled zones in the usually brown subsoil. CI. Pseudogley soils.	
5*.	Soils not having the appearance of gley soils.	6
6.	Strongly acid, nutrient-deficient soils with acid, biologically inert humus formations, strong eluviation of colloidal material, formation of impoverished layers in the top soil and distinct illuvial layers in the top soil and distinct illuvial layers in the subsoil. Occurring generally in heath country or wooded areas with a cool and moist climate.	11
6*.	Soils without acid, biologically inert humus formations.	7
7.	Soils red to brown red coloured, not plastic even with high content, finely dispersed substances, when wet can be neither moulded nor kneaded . . .	8

TABLE 4.13.

Further example of Kubiena's (1953) key number 5 (CD) (p. 194)

Class: steppe soils

Key for the determination of types.

1\. Soil formed from calcareous to highly calcareous parent material. 2

1*. Soil formed from chalk-deficient to chalk-free parent material. 6

2\. Humus horizon extensively decalcified with seam or narrow band-like (*B*) horizon on its lower edge, coloured brown by precipitations of ferric hydroxide.

71\. Degraded chernosem (subtype).

2*. Soil neither with brown seam or band-like (*B*) horizon. 3

3\. Soil occurring only in the desert steppe, always very humus deficient and with low depth of humus, very dry, with open zerophytic plant cover only cultivable under irrigation. 5

3*. Soil usually with a deep, dark-coloured humus horizon, cultivable without irrigation, with closed natural turf cover or used as arable soil. 4

4\. Humus when moist, blackish to dark grey.

XXX. Chernosem.

4*. Humus when moist, chestnut brown.

XXIX. Kastanosem (chestnut-coloured soil).

5\. Humus horizon light grey, often scarcely distinguishable from the mineral soil.

XXVII. Serosem.

5*. Humus horizon grey brown.

XXVIII. Burosem (brown soil of the desert steppe).

6\. Soil occurring only in the desert steppe, always very humus deficient and with low depth of humus, very dry, with an open sparse xerophytic plant cover.

XXXII. Para-serosem.

6*. Soil not occurring only in the desert steppe, generally with deep humus horizon under a closed, natural turf over or used as arable soils

XXXI. Para-chernosem

The soil classification of the 'Soil map of the world' (1974)

The *Soil map of the world* (Dudal 1974) is based on a grouping of soils which has many of the attributes of a classification system. Indeed this grouping of soil entities into binomial classes (with their specified nomenclature) and a key for the allocation of unknown soils to it has been used by a number of workers as a soil-classification system. It has the particular attraction of appearing to be a soil-classification system in a world context and in this it is unique. For convenience we will call it the 'world system'.

The world system was created by a Unesco advisory panel from the very varied maps contributed to the *Soil map of the world*. The soils of the units on these maps were described in terms of diverse soil classes, categories, and nomenclatures, and the panel's job was to reconcile and correlate these into a common gallery of soil classes which would span the whole range of soil occurrence and to define and name them so that they could be used without ambiguity.

The panel created classes at about great group level. These have no categorical name and there is an unnamed subclass level of grouping. For convenience, we will call these world class and world subclass categories.

The 26 world classes comprise fluvisols, gleysols, regosols, lithosols, arenosols, rendzinas, rankers, andosols, vertisols, solonchaks, solonetz, yermosols, xerosols, kastanozems, chernozems, phaeozems, greyzems, cambisols, luvisols, podzoluvisols, podzols, planosols, acrisols, nitosols, ferralosols, and histosols.

Each world class is defined; e.g. the definition for luvisols is

> 'Soils having an argillic horizon which has a base saturation of 50 per cent or more (by NH_4OAc) at least in the lower part of the *B* horizon within 125 cm of the surface; lacking a mollic *A* horizon, lacking an albic *E* horizon overlying a slowly permeable horizon, the distribution pattern of the clay and the tonguing which are diagnostic for planosols, nitosols, and podzoluvisols, respectively; lacking an aridic moisture regime.'

There is also a listing of the 29 local and regional classes that have been merged into this world class (Table 4.14).

The world classes are subdivided into 103 world subclasses. The subclasses are defined and named on deviant profile characteristics which may be intergrade or exgrade in nature. In addition to the orthic variants, there are deviants on 29 variant characteristics which are defined as albic, mollic, gelic etc. and are listed in Table 4.15. The world system then uses a binomial nomenclature, e.g. an albic luvisol is the class luvisol and the subclass albic refers to the presence of a bleached A_2 horizon as defined.

The world system has a key for allocating unknown soils to its classes and subclasses. Table 4.16 and an earlier quotation in Tables 4.14 and 4.15 are taken from Dudal (1974) and are quoted by permission of both U.N.E.S.C.O. and F.A.O. Table 4.16 illustrates the portion referring to yermosols and xerosols.

TABLE 4.14.
Local and regional soil classes included in luvisols for the World soil map

Luvisols include the following:

Grey-brown podzolic	Vertic haploxeralfs
Sols lessivés modaux	Lateritic podzolic soils
Parabraunerde	Red-yellow podzolic soils of high base status
	Sols ferrugineux tropicaux lessivés
Hapludalfs	Grey podzolic soils
Haploxeralfs	Grey wooded soils
Podzolized brown forest soils	Sols podzoliques
Terra rossa	Eutroboralfs
Red-brown earths	Groundwaterlaterites
Terra fusca	Plinthustalfs
Sols fersiallitique lessivés	Plinthoxeralfs
Solos Mediterraneos Vermelhos o Amarelos	
Rhodoxeralfs	Sols lessivés hydromorphes
Haploxeralfs	Sols à gley lessivés
Cinnamonic soils	Podzolic gley soils
Haplustalfs	

TABLE 4.15.
Diagnostic characteristics on which subclasses are distinguished: World soil map

Eutric	Gleyic
Calcaric	Haplic
Dystric	Gypsic
Thionic	Glossic
Mollic	Vertic
Humic	Leptic
Plinthic	Placic
Gelic	Solodic
Cambic	Xanthic
Luvic	Rhodic
Ferralic	Aeric
Albic	Chronic
Ochric	Orthic
Vitric	Takyric
Pellic	

TABLE 4.16.
Portion of key to World soil map classes

Definition and subdivision of XEROSOLS and YERMOSOLS	
Other soils having a weak ochric *A* horizon and an aridic moisture regime; lacking permafrost within 200 cm of the surface	
XEROSOLS (X)	
Xerosols having an argillic *B* horizon; a calcic or gypsic horizon may underlie the *B* horizon	LUVIC XEROSOLS (Xl)
Other xerosols having a gypsic horizon within 125 cm of the surface*	GYPSIC XEROSOLS (Xy)
Other xerosols having a calcic horizon within 125 cm of the surface*	CALCIC XEROSOLS (Xk)
Other xerosols	HAPLIC XEROSOLS (Xh)
Other soils having a very weak ochric *A* horizon and an aridic moisture regime; lacking permafrost within 200 cm of the surface	
YERMOSOLS (Y)	
Yermosols showing takyric features	TAKYRIC YERMOSOLS (Yt)
Other yermosols having an argillic *B* horizon; a calcic or gypsic horizon may underlie the *B* horizon	LUVIC YERMOSOLS (Yl)
Other yermosols having a gypsic horizon within 125 cm of the surface*	GYPSIC YERMOSOLS (Yy)
Other yermosols having a calcic horizon within 125 cm of the surface*	CALCIC YERMOSOLS (Yk)
Other yermosols	HAPLIC YERMOSOLS (Yh)

* The depth requirement varies with the weighted average textural class: less than 125 cm from the surface for coarse textures, less than 90 cm for medium textures, less than 75 cm for fine textures. When both a calcic and gypsic horizons are present the classification of the soil is determined by the diagnostic horizon which occurs nearest to the surface.

It will be noted that the only difference between the xerosols and the yermosols and between their corresponding subclasses is the difference between a 'weak ochric *A* horizon' and a very weak ochric *A* horizon'. The difference between weak and very weak is given as:

'*A very weak ochric A horizon* has a very low content of organic matter with a weighted average percentage of less than 1 per cent in the surface 40 cm if the weighted average sand/clay ratio for this depth is 1 or less; or less than 0.5 per cent organic matter if the weighted sand/clay ratio is 13 or more; for intermediate sand/clay ratios the organic matter content is intermediate. When hard rock, a petrocalcic horizon, a petrogypsic horizon, or a duripan occur between 18 and 40 cm, the contents of organic matter mentioned above are, respectively, less than 1.2 and 0.6 per cent in the surface 18 cm of the soil.

A weak ochric A horizon has a content of organic matter which is intermediate between that of the very weak ochric *A* horizon and that required for the mollic *A* horizon.'

The authors of the world system state that different standards of generalization in the original soil classes made it impossible for them to produce a hierarchical system, hence their system is not a taxonomy but a 'monocategorical classification'. They claim that, since the diagnostic horizons were formulated for their significance to soil-forming processes, the soil classes have predictive value for the use of the soil (loc. cit., p. 10).

Comment. This system is similar to other soil-classification systems. The category of world classes bears most of the attributes of a gallery of conceptual classes, which we have found to be a feature of all systems except those of Northcote, Muir, and MacVicar and de Villiers.

The world classes are formed by a mental generalization of many similar classes coming from many different parts of the world. For instance, the luvisol class (see Table 4.14) is a generalization of soil classes from 29 different regions. This is not an impossible generalization, but the total associated variance in the final class coming to it from all of the 29 regions must be so broad that use of the class for predicting use values must be very poor, except for those values which may reside in the keying characteristics themselves. The authors claim that, because the chosen differentiae represent soil-forming processes, the predictive value of the classes for use of the soil is high; this, however, is a *non sequitur* as the edaphic and genetic factors of a soil are not the same and are not necessarily similar or associated.

The world classes and subclasses are defined by a very few precisely defined explicit soil characteristics, and they are keyed for identification by an even sparser set of criteria, these being confined in the main to morphological characteristics of the profile. The example of the xerosols and yermosols is quoted above. The success of the classification at predicting useful soil characteristics will depend on the strength of the patterns of association among the characteristics within the classes, but there is no evidence that this has been analyzed. Instead we find that a great deal of effort has gone into making precise definitions and keys on the assumption that this puts precision into the associated characteristics of the system. However, if there is much diverse variation in the data, a key will be ineffective in reducing it and the classes produced will be 'key-only classes' containing a great amount of 'taxonomic noise'. Since diverse variation is at a maximum on the world scale, we must conclude that a world system has maximum taxonomic noise in its classes and that these may be of only limited use for predictions.

Soil classification for England and Wales (Avery 1973)

This classification system is fairly typical of modern productions. It is based on a gallery of 10 conceptual classes, and two lower categories are formed by a rationalized elaboration of these. The classes are described and defined by reference to diagnostic horizons and defined soil materials, and these criteria serve both as a bifurcating key and as elements in the definition of classes. The system operates entirely on objective profile characteristics. There are 10 *major groups* listed in Table 4.17.

TABLE 4.17.
Major groups of the classification of England and Wales (Avery 1973)

Terrestrial raw soils	Podzolic soils
Hydric raw soils	Surface-water gley soils
Lithomorphic soils	Groundwater gley soils
Pelosols	Man-made soils
Brown soils	Peat soils

The major groups are subdivided into 43 groups on the basis of diagnostic horizons and material composition, and the groups are subdivided into 109 subgroups on the basis of diagnostic horizons and soil materials.

The rationalized breakdown into groups and subgroups is illustrated for the major group of brown soils in Table 4.18.

Series. This is both the class formed by subdividing subgroups on the basis of texture, profile contrast, origin, and mineralogy of the material, and the mapping unit used in detailed soil survey.

The system includes a list of definitions of diagnostic horizons and other characteristic soil materials. There is no separate key to the system, but the purpose of a key is served by the definitions of horizons and materials and a degree of precision is attained.

The stated purpose of the soil-classification system is to provide a basis for the legends of soil maps in England and Wales (Avery 1973, p. 325). The possibility that a problem may arise in reconciling the soil-map units and the taxa of the classification system is recognized, but the author does not see that this will necessitate serious compromise over boundaries. This is a general problem in soil classification for soil survey.

This classification is specifically intended to be relevant to problems in land use, working through the intermediary of the soil map. The author recognizes the limitations in this respect brought about because it is based on profile characteristics alone and excludes site and climatic factors, and he considers that it may be desirable to subdivide soil series as mapped into slope phases and eventually perhaps into climatic phases too.

Comment. Like other national and regional systems the English system is as unique as the particular realm of experience it is based on. It covers the range of soils from podzols to brown soils with secondary themes of wetness and peatiness. Its structure conforms to what has become a fairly standard pattern. It probably has a rather lower level of precision, particularly in its definitions and keying operations, than *Soil taxonomy*. Its hierarchical form is less pronounced and it has no formal keys. These may cause a little ambiguity in allocating new soils to it, but they will lessen the compromise involved in reconciling taxonomic units with soil areas that have boundaries with marked external expression in the field.

The English system may not relate as well to agricultural usage as the United States' system because it lacks criteria of climatic moisture regime and temperature regime. However, each system matches the need of its country in a general way. Moisture deficit is

TABLE 4.18.
Subdivision of the major group of brown soils in the soil classification of England and Wales (Avery 1973)

Major group	Group	Subgroup
5 *Brown soils* Soils, excluding pelosols, with weathered, argillic or paleoargillic *B* and no diagnostic gleyed horizon at 40 cm or less	5.1 *Brown calcareous earths* Non-alluvial loamy or clayey without argillic horizon	5.11 Typical brown calcareous earth 5.12 Gleyic brown calcareous earth 5.13 Stagnogleyic brown calcareous earth
	5.2 *Brown calcareous sands* Non-alluvial, sandy, without argillic horizon	5.21 Typical brown calcareous sand 5.22 Gleyic brown calcareous sand
	5.3 *Brown calcareous alluvial soils* In recent alluvium	5.31 Typical brown calcareous alluvial soil 5.32 Gleyic brown calcareous alluvial soil
	5.4 *Brown earths (sensu stricto)* Non-alluvial, non-calcareous, loamy or clayey, without argillic horizon	5.41 Typical brown earth 5.42 Stagnogleyic brown earth 5.43 Gleyic brown earth 5.44 Ferritic brown earth 5.45 Stagnogleyic ferritic brown earth
	5.5 *Brown sands* Non-alluvial, sandy or sandy gravelly	5.51 Typical brown sand 5.52 Gleyic brown sand 5.53 Stagnogleyic brown sand 5.54 Argillic brown sand 5.55 Gleyic argillic brown sand
	5.6 *Brown alluvial soils* Non-calcareous, in recent alluvium	5.61 Typical brown alluvial soil 5.62 Gleyic brown alluvial soil
	5.7 *Argillic brown earths* Loamy or clayey, with ordinary argillic *B*	5.71 Typical argillic brown earth 5.72 Stagnogleyic argillic brown earth 5.73 Gleyic argillic brown earth
	5.8 *Paleo-argillic brown earths* Loamy or clayey, with paleoargillic *B*	5.81 Typical paleoargillic brown earth 5.82 Stagnogleyic paleoargillic brown earth

less of a problem in England than in the United States and the temperature range is less. An English system should be more concerned with drainage differentiae than the US system and this one is.

The Canadian system (Anon 1974)

The Canadian soil classification is interesting in its contrasts with its counterpart in the United States. The two countries have many confluent climatic and ecological zones in their 4200 km of common border, and they are at comparable levels of technology. Physical grounds for diversity in their soil-classification systems should be at a minimum.

The structure of the Canadian system is based on a gallery of eight orders, and these are subdivided by rational elaboration into two lower categorical levels—the great groups and subgroups. These are followed in turn at the level of soil-survey involvement by family, series, and phase classes. An outline of the orders and great groups is presented in Table 4.19.

TABLE 4.19.
Orders and great groups of the Canadian system

Orders (8)		
	Chernozemic	Brunisolic
	Solonetzic	Regosolic
	Luvisolic	Gleysolic
	Podzolic	Organic
Great Groups (23)		
		Melanic brunisol
	Brown	Eutric brunisol
	Dark brown	Sombric brunisol
	Black	Dystric brunisol
	Dark grey	Regosol
	Solonetz	Humic gleysol
	Solod	Gleysol
	Grey-brown luvisol	Eluviated gleysol
	Grey luvisol	Fibrisol
	Humic podzol	Mesisol
	Ferrohumic podzol	Humisol
	Humoferric podzol	Folisol

The orders are subdivided into 23 great groups and the groups into 165 *subgroups* on the basis of deviation, intergrade or exgrade, defined diagnostic horizons, and other defined profile characteristics such as colour, structure, and texture. Table 4.20 illustrates the subdivision of the chernozemic order into subgroups. It is quoted from Anon (1974) by permission of the Minister of Supply and Services, Canada.

The 800–1000 *families* are groupings of the series within a subgroup according to texture, texture contrast, mineralogy, depth, pedoclimate, temperature class, and moisture class. The *series* (about 3000) is a soil body in the field, of limited defined variation in texture, hue, and consistence.

The system is supplied with a comprehensive catalogue of profile descriptions and definitions within the framework of *A*, *B*, and *C* horizons.

There is no soil key as such, but the orderly definition and arrangement of class definitions serves the purpose of a key without being quite so narrow in its elections. The differentiae of classes are embedded in their definitions and are mainly morphological attributes such as colour, structure, texture, and arrangement of horizons into profiles. Temperature and humidity regimes are not employed as differentiae above the family level, but they are part of the family definition and name.

Comment. The Canadian system is fairly straightforward with a gallery of conceptual classes leading into several categories of rationalized elaborations. There is some ambivalence in the role of conceptual classes, some being orders and some great groups, and this probably results from unequal rates of growth in different parts of the system. The structure of the Canadian system is similar to that in the United States *Soil taxonomy* except that it has one category fewer. It uses the series–polypedon relationship, and in Canada too the family seems to be a separate, even a duplicated, edaphic grouping. The Canadian system has no explicit key, and its use of horizon and profile definitions for the keying function makes the allocation of profiles to classes rather less arbitrary.

In discussing (loc. cit., p. 183) the prospect of correlating their classification system with that of the United States, the Canadians point out that two alignments are needed for correlation; first there must be identical criteria for the description and definition of soil

TABLE 4.20.
Subdivision of the chernozemic order in the Canadian system (Anon 1974)

Order	Great group		Subgroup
Chernozemic	1.1 Brown	1.11	Orthic brown
		1.12	Regobrown
		1.13	Calcareous brown
		1.14	Eluviated brown
		1.11-2.11	Solonetzic brown
		1.14-2.21	Solodic brown
		1.1-/5	Saline brown
		1.1-/6	Carbonated brown
		1.1-/7	Grumic brown
		1.1-/8	Gleyed brown
		1.1-/9	Lithic brown
	1.2 Dark brown	1.21	Orthic dark brown
		1.22	Rego dark brown
		1.23	Calcareous dark brown
		1.24	Eluviated dark brown
		1.21-2.11	Solonetzic dark brown
		1.24-2.21	Solodic dark brown
		1.2-/5	Saline dark brown
		1.2-/6	Carbonated dark brown
		1.2-/7	Grumic dark brown
		1.2-/8	Gleyed dark brown
		1.2-/9	Lithic dark brown
	1.3 Black	1.31	Orthic black
		1.32	Rego black
		1.33	Calcareous black
		1.34	Eluviated black
		1.31-2.12	Solonetzic black
		1.34-2.22	Solodic black
		1.3-/5	Saline black
		1.3-/6	Carbonated black
		1.3-/7	Grumic black
		1.3-/8	Gleyed black
		1.3-/9	Lithic black
	1.4 Dark gray	1.41	Orthic dark gray
		1.42	Rego dark gray
		1.43	Calcareous dark gray
		1.41-2.12	Solonetzic dark gray
		1.41-2.22	Solodic dark gray
		1.4-/5	Saline dark gray
		1.4-/6	Carbonated dark gray
		1.4-/7	Grumic dark gray
		1.4-/8	Gleyed dark gray
		1.4-/9	Lithic dark gray

TABLE 4.21.
Nearest apparent orders in the Canadian and United States classification systems

Canadian	United States	
Chernozemic	Mollisols	
Solonetzic		
Luvisolic	Alfisols	
Podzolic	Spodosols	
Brunisolic	Inceptisols	
Gleysolic		
Organic	Histosols	
Regosolic	Entisols	
		Oxisols
		Vertisols
		Aridisols
		Ultisols

profile and diagnostic terms, and secondly there must be identical galleries and categories of classes. In making correlations between classification systems we are inclined to think only of the need for reconciling the arrays of classes. But the first alignment is a prerequisite, and this is not met in the Canadian-United States correlation. Alignment is not only a case of abandoning one set of descriptions and definitions for another, but also of abandoning a system which fits the modes and frequency distributions of one geographic region for a system which was developed to fit the modes and frequency distributions of some other region.

The Canadian system differs noticeably from the United States system in its classes at the order level. The orders of both systems are listed against their nearest parallel class in Table 4.21.

Several orders occur in the United States which do not occur in Canada, or occur to such small extent that they are not rated of first importance in classification. This applies to oxisols, vertisols, ultisols, and aridisols; the last is present in lower categorical levels in the Canadian system and defined in other terms. These differences can be readily associated with the differences in geography of the two countries. Similarly, the greater attention to solonetzic and gleyed soil features in Canada can be associated with differences of geography and is more one of emphasis than of principle. But there is the additional difference in definition and probably differences brought about by different institutional history in the two countries. There is plenty of opportunity for subtle differences of a philosophic and personal character, as well as of a geographic kind, to become involved in the conceptual classes and in the associated fabric of definition of classification systems. Such is the nature of classification that, even with very similar technological and scientific levels and a similarity in geography and history, there is still value in keeping the U.S. and Canadian classification systems separate even though there are the expected advantages, in predictive terms, of merging the systems.

The 1938 U.S.D.A. soil classification (Baldwin, Kellog, and Thorp 1938)

We include a reference to this classification because it marks an important stage in the history of development of thought on soil classification.

The main part of the 1938 classification is its gallery of 36 great soil groups. These comprise the ideal form of a 'gallery of conceptual classes'; they are centroid complex concepts comprising soil profile, soil environment, and pedogenetic attributes. There are no rationalized boundaries between classes and, although there is some rationalized elaboration of classes both upward and downward, these developments were of little significance and some still remain to be established.

The system was planned to be hierarchical and it extends from the soil type of the soil surveyor upwards through four categories to the order. There is no key, however, nor any sets of definitions of class boundaries. All classes are described by their central concepts and their descriptions are very brief.

The 1938 system begins with the zonal–intrazonal division which relates to the concept of 'normal' and 'non-normal' soils and has since been dropped by general consent. A general outline of the system is given in Table 4.22. This is copied from Baldwin *et al.* (1938) by kind permission of the Chief Education and Publication Branch of the Soil Conservation Service U.S.D.A.

The categories below the great soil groups are families and then there are series and types which are generated by detailed soil surveys and not created by the system. It was conceded that the families were for the future, but it seems clear that they were to be combinations of series rather than subdivisions of great soil groups. The series, as a category, is an independent point of origin of the classification and the

TABLE 4.22.
The classification of Baldwin et al. (1938)

	Order	Suborder	Great soil group
Zonal soils	Pedocals	Soils of the cold zone	Tundra soils
		Light coloured soils of arid regions	Desert soils Red desert soils Sierozem soils Brown soils Reddish brown soils
		Dark coloured soils of arid sub humid and humid grasslands	Chestnut soils Reddish chestnut soils Chernozem soils Prairie soils Reddish prairie soils
	Pedalfers	Soils of the forest grassland transition	Degraded chernozem soils Non-calcic brown soils
		Light coloured podzolised soils of the timbered regions	Podzol soils Brown podzolic soils Gray-brown podzolic soils
		Lateritic soils of forested warm-temperate and tropical regions	Yellow podzolic soils Red podzolic soils (and terra rossa) Yellowish-brown lateritic soils Reddish-brown lateritic Laterite soils
Intrazonal soils		Halomorphic (saline and alkaline) soils of imperfectly drained arid regions and littoral deposits	Solonchak soils Solonetz soils Soloth soils
		Hydromorphic soils of marshes, swamps, seep areas and flats	Wiesenboden Alpine meadow soils Bog soils Half bog soils Planosols Ground water podzols Ground water laterite soils
		Calcimorphic	Brown forest soils (braunerde) Rendzina
Azonal soils			Lithosols Alluvial soils Sands (dry)

beginning of an upward rationalization into what was intended to become several categories.

Comment. This classification is a good representation of current thinking at the time of its origin. Many of its differentiae are environmental and not soil characteristics; other criteria refer to central concepts and not to class boundaries, and no guidance at all was given on many of the distinctions that need to be made. It would be difficult to use it to identify an unknown soil.

De Sigmond's soil classification of Hungary (1938)

This system represents an important stage in turning away from environmental characteristics to characteristics of the soil itself.

De Sigmond's system is hierarchical with six categories—main groups, subgroups, soil orders, main types, subtypes, and local varieties. The first three categories are shown in Table 4.23 which is copied from de Sigmond (1938) by kind permission of George Allen & Unwin.

Table 4.24 presents the subdivisions of the soil order for calcium soils as an example of the categories below soil orders.

The system has no key for the identification of new soils and identification would need to be effected by a series of matching operations at each categorical level. The description of the classes varies; those for main groups and subgroups are based on the dynamic character of the material composing the soil while soil orders are distinguished by various specific features of the soil dynamics. De Sigmond follows Gedroiz in giving primary importance to the genetic and dynamic characteristics of soils, which he considers are expressed in the kind and amount of the soil organic matter and the exchange characteristics and cation status of its exchange complex. He chooses not to use the environmental characteristics as pedogenetic factors but to use the dynamic characteristics of the soil which reflect them.

The main types are not derived from higher categories but are centroid concepts of 'whole-soil' classes from direct observations of soil distribution in nature. Correspondingly, the descriptions of the main types include full descriptions of the soil profile, vegetation association, and the environment in addition to the dynamics of the profile. The lower categories (subtypes and local varieties) again revert to logical subdivisions of the data, in this case of the data brought in as the main types.

TABLE 4.23.
De Sigmond's (1938) first three categories

Main groups	Subgroups	Soil orders
1. Organic soils	1. Raw organic soils i.e. turfy soils	1. Turfs, poor in bases
		2. Turfs, rich in bases but not salty
		3. Salty turfs
	2. Humified organic soils, i.e. peat soils	4. Acid peat soils
		5. Neutral peat soils
		6. Salty peat soils
2. Organic mineral soils	3. Raw organic mineral soils	7. Endoadynamic soils
		8. Ectoadynamic soils
		9. Pseudoadynamic soils
	4. Humic siallites	10. Hydrogen soils
		11. Calcium soils
		12. Sodium soils
	5. Ferric siallites	13. Brown earth
		14. Red earth
		15. Yellow earth
	6. Allites	16. Pure allites
		17. Siallitic allites
		18. Bauxitic allites
3. Purely mineral soils	7. Raw mineral soils	19. Soils of mixed rock debris
		20. Soils of mineral grits
		21. Soils of fine mineral dust
	8. Mineral soils with some decomposition	22. The calcium of the mineral matter partly mobilized
		23. The silica of the mineral matter partly mobilized
	9. Mineral soils with end products of decomposition	24. Soils with easily soluble salt crust
		25. Soils with slowly soluble salt crust

Comment. This system was innovatory in its day because the criteria on which it operates are all soil characteristics, not external factors. But in spite of this it must have proved very difficult to allocate a soil entity to its class because most of the classes are defined on their central concepts without class boundaries.

Even in this early classification system the general structure of classification systems which we have postulated can be discerned. A gallery of conceptual classes—the main types—is central to the

TABLE 4.24.

Soil order – 11. calcium soils (de Sigmond 1938)
Subdivision to main types and subtypes

Main types 1 – Tshernozem or black (dark brown) steppe soils

Subtypes:	*Local varieties*
(1) With excess leaching (2) With medium leaching (3) With slight leaching	Local orographical variants Local hydographical variants Local climatic variants Local parent material variants Local age variants

Main types 2 – Prairie soils

Main type 3 – Rendzinas

Subtypes:

(1) Carbonate rendzinas
(2) Sulphate rendzinas

Main type 4 – Tropical and subtropical black earths (black soils)

Main type 5 – Brown steppe soils

Subtypes:

(1) Chestnut steppe
(2) Light-brown steppe

Main type 6 – Grey steppe soils

Main type 7 – Secondary and tertiary calcium soils

Main type 8 – Calvero soils (light chestnut, dry forest soils)

system. These are based on the general observation of a wide range of soils. This gallery was then subject to rationalized elaboration, downward by splitting off minor deviant themes in the order that emerges and upward into increasingly broad generalizations that seem to marshal the data best. In de Sigmond's system this upward elaboration is given much more significance than it generally has in other systems. De Sigmond considered that the organic matter and the exchange complex are the integral characteristics of soil dynamics and that, because there is an orderly subdivision of the range of these characteristics in the higher categories (main groups, subgroups and soil orders), the system has a unique relevance which others lack.

It is hard to agree with this claim because the higher categories do not form or generate the main types which are independent intuitive creations formed on the basis of widespread observation of soils. The lower categories (subtypes and local varieties) are truly formed by subdividing the main types, so these three categories together comprise a system which can operate without the three higher categories. The higher categories are an independent creation, albeit a valuable comment on the essential local range of soil character. This divorce of the upper categories is not brought about because of the correctness or otherwise of de Sigmond's claim for the importance of their criteria; it occurs because they do not form or define the lower categories, the main types and subtypes.

J. Muir's system (1969)

This system is included here because it carries several trends in classification to the extreme of their development. The system is a key which has been developed around a theory of soil genesis. The theory of soil genesis presented in Muir's Principle of Developmental Sequences postulates that the essential *sequence* of soil-profile development (losses from horizons and gains to horizons), passing from least to most developed, is

(a) water-soluble salts
(b) sparingly soluble salts
(c) iron/humus
(d) iron oxide
(e) silica
(f) clay.

He then classifies horizons in the terms of Table 4.25 and defines them in terms of one or two morphologic properties.

Then five categories of soil classes are defined on the basis of horizon groupings as illustrated in Table 4.26.

All of the classes are defined on precise cut-offs in morphologic or (a few) analytic criteria These criteria are chosen to represent the essential soil dynamics, and this aspect is given prime importance by Muir (1962). He writes [loc. cit., pp. 24, 25] 'Before the systematist can order objects by all properties he must have a principle or theory concerning the objects so that their properties can be given an order or importance'; and 'The better the theory the more the essential nature of the object is understood and incorporated into a system of classification'; and [p. 27] 'A proper understanding of the processes and factors provides the guidance required to construct a natural

TABLE 4.25.

Genetic processes to be identified in soil horizons
Muir (1969)

A horizons	*E* horizons	*B* horizons	*C* horizons
Basiallic	Lixiviated	Salic	Salic
Lixiviated	Solodised	Calcic	Calcic
Solodised	Leached	Argillic	Basiallic
Leached	Lessivated	Humoferric	Argillic
Lessivated	Desilicated	Ferric	Ferric
Desilicated	Podzolised		Desilicated
Podzolised			Podzolised
Salic			
Calcic			
Argillic			

TABLE 4.26.

The main subdivisions of Muir's (1969) classification

Orders – salisols, calcisols, basidisols, acidisols, organosols

Suborders (1) regosalic, salic, lixiviated, solidized
(2) carbonate, calcic, leached
(3) basiallic, regillic, argillic, lessivated
(4) redesilicated, desilicated, podzolized

Divisions – each suborder ⟨ aerasols / gleysols

As an example of lower categories:

Major soil group – brown siasols
Subgroups – sandy brosiasols, loamy brosiasols, clayey brosiasols
Major soil group – orthic siagleys
Subgroups – sandy orsiagleys, loamy orsiagleys, clayey orsiagleys
Major soil group – vadose siagleys
Subgroups – loamy varsiagleys, clayey varsiagleys, frago varsiagleys

system based on properties, but neither processes nor factors can take the place of properties.'

Throughout Muir's system the evidence of leaching or deposition in the profile of bases, silica, and sesquioxides are given importance, as is the role of the exchange complex in both its organic and mineral components. Muir considers that these elements comprise the 'proper understanding of the processes' of soils and his categories and classes are designed within this framework. He has used a sparse selection of soil characteristics, most frequently morphological, to represent the soil-process conditions that he wants to highlight, and these serve as a key.

Comment. While many people would agree that the exchange complex, its characteristics, and the ions associated with it express the fundamental nature of a soil, it may not follow that groupings on this basis will give the classification system the most useful and informative classes. It would be better if Muir had tested the orderliness of other characteristics in classes based on his criteria rather than rely on the authority of J. S. Mill. Muir's stance of expecting a perfect classification because it is based on correct theory amounts to a claim of perfect orderliness in his soil-data pool, and this cannot be presumed. Not only does he count on general orderliness but, in particular, on high associations between the morphologic features (e.g. hue, chroma, value) of his key and basic soil processes. It is difficult to sustain the hypothesis that a correct theory gives a correct classification when we see that so many classifications have diverse or no theory and yet they all seem to work as well or as badly as each other.

The South African soil classification

The South African system (Macvicar and de Villiers 1977) is a renewed approach to soil classification which is intended to correct a lack of rigour in definition and differentiation, and a lack of comprehensiveness in earlier classifications.

The system comprises two categories—soil forms (of which there are 41) and soil series (of which there are 504).

The soil forms are defined by reference to five surface horizons and 15 subsurface diagnostic horizons. The diagnostic horizons have descriptive names and are defined by two or three attributes which serve as keying properties during the identification of soil entities.

Attributes used as differentiae between diagnostic horizons are

Surface horizons: per cent organic matter, exchange cation status, structure, per cent clay etc.

Subsoil horizons: colour, structure, material, E horizon, G horizon etc.

A soil form is defined and keyed by its possession of prescribed surface and subsurface diagnostic horizons, but the system does not describe it or its site and land-use aspects any further. It has a local geographic name derived from a well known soil series which belongs to it. The presence of *E* (bleached) horizon receives high significance in the system. Seven soil forms have the same overlying and underlying horizons but differ from seven other soil forms in the presence of a diagnostic *E* horizon.

Soil series are defined as profiles belonging to a soil form which, in addition, comply with one or two specified differentiae such as per cent clay, grade of sand fraction, calcareousness, base saturation, pH, colour, etc.

As usual, the soil series have locality names. They are given the minimum definition for their identification, but they are not described either as profiles or as mappable areas of land.

The use of the South African system for identifying a soil profile (Macvicar and de Villiers loc. cit., p. 5) involves entry at the soil-form level by using the differentiae of the diagnostic horizons and then checking for the soil-series differentiae. From its character as a bi-categorical system, it leads to a binomial system of nomenclature —a soil-form name followed by a soil-series name. However, each soil series belongs to one soil form only and all soil series are already defined, so the binomial use here does not involve the degree of freedom that binomial use brings to botanical classification.

Comment. The South African system as published is a key for identifying soils, and its diagnostic horizons contain the differentiae whereby an unknown soil is allotted to its soil form and its soil series.

The system illustrates well the drastic change in emphasis which may occur once a classifcation system is created and then comes into use. It was formed after many years of soil survey had accumulated many soil series, each of which had been formed as a conceptual class to fit a particular soil-survey project. These series as experienced by the soil surveyor then moulded the keying system now proposed which presumably now offers no differentiae for which past experience has presented no need. But any future series must fit the system, which thereby has become a narrowly defined keying system. This is its most undesirable feature. The success of the system at predicting the associated attributes of any future classes will depend on whether or not the soil-survey experience of the past serves as a satisfactory representative soil-data pool for future soil-survey projects. There is no reason to assume that this will be so. Even though both realms of experience are within South Africa we must expect it to fail at times. The form of its failure will be that associations of classes with important attributes will sometimes be found lacking, and at times modal distributions will be split between two classes. Probably no national classification system can avoid these failures altogether.

It seems unnecessary to add any general comment to those above. The great variety in the approaches to soil classification serve to make the point that we have as yet no general agreement on what should be its basis or its objectives.

References

Anon (1974). *The system of soil classification for Canada*. Publication No. 1455, Canadian Department of Agriculture, Ottawa.

Avery, B. W. (1973). Soil classification in the Soil Survey of England and Wales. *J. Soil Sci.* **24**, 324–38.

Baldwin, M., Kellogg, C. E., and Throp, J. (1938). Soil classification. In *Soils and Men. Yearbook of Agriculture 1938*, pp. 979–1001. Washington, D.C.

Basinsky, J. J. (1959). The Russian approach to soil classification and its recent development. *J. Soil Sci.* **10**, 14–26.

Blackburn, G. (1960). Putting Russian soils on the map. *Soils and Fert.* **23**, 155–8.

Bleeker, P. and Speight, J. G. (1978). Soil-landform relationships at two localities in Papua New Guinea. *Geoderma* **21**, 183–98.

Butler, B. E., Baldwin, J. G., Penman, F. and Downes, R. G. (1942). Soil survey of part of County Moira, Victoria. *C.S.I.R. Bul.* 152.

Christian, C. S. and Stewart, G. A. (1968). Methodology of integrated studies. *Proc. Toulouse Conf. 1964*, pp. 233–80. UNESCO, Paris.

Clarke, G. R. and Beckett, P. (1971). *The study of soil in the field.* Clarendon Press, Oxford.

Cline, M. G. (1963). Logic of the new system of soil classification. *Soil Sci.* **96**, 17–22.

Colwell, J. D. and Esdaile, R. J. (1968). The calibration, interpretation and evaluation of tests for the phosphorus fertilizer requirements of wheat in northern N.S.W. *Aust. J. Soil Res.* **6**, 105–20.

Dudal, R. (1974). *Soil map of the world*, Vol. 1. UNESCO, Paris.

FitzPatrick, E. A. (1971). *Pedology, a systematic approach to soil science.* Oliver & Boyd, Edinburgh.

Gerasimov, I. P. and Glazovskaya, M. A. (1965). *Fundamentals of soil science and soil geography.* Israel Program for Scientific Translations, Jerusalem.

Gilmour, J. S. L. (1962). *Introduction. Classification: an inter-disciplinary problem.* Aslib Proceedings **14**, 223–5.

Howard, A. and McIntyre, G. A. (1943). A survey, census and statistical study of the horticultural planting in M.I.A. New South Wales. *C.S.I.R. Bul.* 168.

Johnson, W. M. (1963). The pedon and the polypedon. *Soil Sci. Soc. Am. Proc.* **27**, 212–15.

Jones, T. A. (1959). Soil classification—a destructive criticism. *J. Soil Sci.* **10**, 196–200.

Knox, E. G. (1965). Soil individuals and soil classification. *Soil Sci. Soc. Am. Proc.* **29**, 79–84.

Kubiena, W. L. (1953). *The soils of Europe.* Thomas Murby, London.

Leeper, G. W. (1954). The classification of soils—an Australian approach. *Trans. 5th Intl. Congr. Soil Sci.* (Leopoldville) Vol. IV, 217–26.

Loveday, J. (1964). Study of the relationship between yield of irrigated lucerne and the properties of some grey and brown soils of heavy texture in south-west New South Wales. *Aust. J. Soil Res.* **2**, 96–110.

Loveday, J. and McIntyre, D. S. (1966). Soil properties influencing growth of subterranean clover in the Coleambally irrigation area N.S.W. *Aust. J. Exp. Agric. Anim. Husb.* **6**, 283.

Macvicar, C. N. (1969). A basis for the classification of soils. *J. Soil Sci.* **20**, 141.

—— and de Villiers, J. M. (1977). Soil classification—a binomial system for South Africa. *Soil & Irrig. Res. Inst. Dept. Agric. Tech. Serv. Sci. Bull.* 390.

Muir, J. W. (1962). The general principles of classification with reference to soils. *J. Soil Sci.* **13**, 22–30.

—— (1969). A natural system of soil classification *J. Soil Sci.* **20**, 153–66.

Northcote, K. H. (1960). A factual key for the recognition of Australian soils. *C.S.I.R.O. (Aust.) Soil Div. Divl. Rep.* 4/60.

—— (1971). *A factual key to the recognition of Australian soils* (3rd edn). Rellim Technical Publications, Glenside, South Australia.

—— Hubble, G. D., Isbell, R. F., Thompson, C. H. and Bettenay, E. (1975). *A description of Australian soils.* C.S.I.R.O., Melbourne.

de Sigmond, A. A. J. (1938). *Principles of soil science.* Thomas Murby, London.

Smith, G. D. (1963). Objectives and basic assumptions of the new soil classification system. *Soil Sci.* **96**, 6–16.

Soil Survey Staff (1951). *Soil survey manual.* U.S. Dept. Agriculture Handbook No. 18, Washington, D.C.

—— (1975). *Soil taxonomy.* U.S. Dept. Agriculture Handbook No. 436, Washington, D.C.

Sokal, R. R. and Rohlf, F. J. (1969). *Biometry.* Freeman, San Francisco.

Stace, H. C. T., Hubble, G. D., Brewer, R., Northcote, K. H., Sleeman, J. R., Mulcahy, M. J. and Hallsworth, E. G. (1968). *A handbook of Australian soils.* Rellim Technical Publications, Glenside, South Australia.

Stephens, C. G. (1953). *Manual of Australian soils.* C.S.I.R.O. Melbourne.

Webster, R. (1968). Fundamental objections to the 7th approximations. *J. Soil Sci.* **19**, 354–66.

—— (1977). *Quantitative and numerical methods in soil classification and survey.* Clarendon Press, Oxford.

—— and Butler, B. E. (1976). Soil classification and survey studies at Ginninderra. *Aust. J. Soil Res.* **14**, 1–26.

Wilding, L. P., Jones, R. B. and Schafer, G. M. (1965). Variation of soil morphological properties within Miami, Celina and Crosby mapping units in Western Central Iowa. *Soil Sci. Soc. Am. Proc.* **29**, 711–17.

Author index

Subject index